4

Oil, The Middle East
And The World

4

THE WASHINGTON PAPERS

Oil, The Middle East
And The World

Charles Issawi

The Center for Strategic and International Studies
Georgetown University, Washington, D.C.

THE LIBRARY PRESS
NEW YORK
1972

CONTENTS

Introduction

INTRODUCTION

If an epoch is to be designated by its most essential material, ours will have to be called either the Paper or the Oil Age. Oil has become by far the leading fuel and probably the most indispensable single raw material of contemporary industrial civilization. It supplies half the world's energy consumption and is the mainstay of industry, the lifeblood of transport and the sinews of war. At the same time, the oil industry has a peculiar and completely unprecedented feature: the whole Western world, and more particularly its two most dynamic segments, Western Europe and Japan, is greatly dependent for its essential oil supplies on a small number of underdeveloped countries lying outside its borders, mainly in the Middle East and North Africa. The nature and extent of this dependence, and the main consequences that flow from it, form the subject of this study.

Chapter One examines the world's energy pattern — the production, consumption, and resources of the main fuels, with projections to 1980. Chapter Two studies the economic factors that account for the predominance of Middle Eastern oil, and also the financial gains derived from it by the industry and the governments of the producing countries. In Chapter Three the political forces affecting the oil industry are analyzed, both those arising within the Middle East and those impinging on it from outside. Lastly, Chapter Four takes up the various measures that have been suggested to reduce the risks arising from the great dependence of the West on Middle Eastern oil, and suggests various policies which may decrease, even if they cannot eliminate, the consequent dangers.

I should like to thank Mohammed Yeganeh, J. C. Hurewitz, my wife, whose critical comments were most helpful, and Mrs. C. Wheeler, who typed the by no means clear manuscript.

C.I.

CHAPTER ONE

The Energy Pattern

E conomic activity, or more precisely the production and transport of goods, has always required relatively large amounts of energy. For thousands of years this was provided exclusively by human or animal power, with the important exception of sailing ships. And it may be noted in passing that until the diffusion of breast harnessing, in the early Middle Ages, massed human labor was much more effective than massed animal power, and was therefore generally used for transporting and raising very heavy construction materials. An important step forward was the invention of the water mill, shortly before the Christian Era, and of the windmill, probably in the seventh century A.D. These made it possible not only to tap inanimate sources of energy but also to secure greater amounts of concentrated power: whereas the horsepower of a man is 0.1 or below, that of a water mill is 2-5 and that of a post windmill 2-8.[1] Nevertheless, it has been estimated that until the Industrial Revolution some 80-85 per cent of total energy was provided by plants, animals and men.[2]

The next long step was taken with the introduction of the steam engine in the eighteenth century. Although the early engines were grossly inefficient, compared to either animal converters of energy or water mills (even that of Watt's was below 5 per cent – against 40 per cent at present) they could produce as much as 100 h.p. And although waterpower continued to provide the greater part of the energy used by industry in Britain, the United States, and Western Europe during the first decades of the nineteenth century, steam power forged ahead and by the middle of the century had established its dominance. The aggregate horsepower of steam engines in Britain in 1800 may have been about 10,000; by 1840 the figure was over 600,000, and by 1900 over 14,000,000. Corresponding estimates for the whole world are about 1,650,000 h.p. in 1840 and over 70,000,000 in 1900.[3] This naturally implies a huge increase in the amount of coal produced since coal began to displace wood as the main fuel. In Britain, about 10 million tons were mined in 1800, or 1.1 tons per head of population, but by 1860 the total had risen to 80 million tons, or 3.5 per head, and by 1910 to 264 million, or 6.1 per head. Corresponding figures for the world were about 15 million, 132 million and 1,057 million tons, and at the outbreak of the First World War, coal and lignite constituted over 90 per cent of the world's commercial sources of energy. The latter grew at a compound rate of about 3 per cent a year between 1860 and 1913.

Compared to coal, petroleum was a latecomer, and assumed importance as a source of energy only after the First World War. Commercial production started in Rumania, the United States and elsewhere, in the 1850s, but total world output reached the 1,000,000 ton (7 million barrels) level only in 1872, and the 10,000,000 level in 1890, while the 100,000,000 ton (766 million barrels) mark was not passed until 1921. Correspondingly, its share of world production of commercial sources of energy rose to 6 per cent in 1913 and 10 per cent shortly after the First World War.

Trends 1925-1970

Developments since 1925 are shown in Table 1. The figures in the table may be supplemented by further information derived from the same source. First, over the whole period, total energy consumption rose at a compound rate of 3.4 per cent a year; a breakdown into subperiods shows wide fluctuations, the rate for 1925-1950 being only 2.3 per cent and that for 1950-1968 about 5 per cent.

TABLE 1

World Energy Consumption
(million metric tons coal equivalent)

	1925	1938	1950	1960	1968
Solid fuels	1,230	1,292	1,593	1,999	2,315
Liquid fuels	197	378	722	1,499	2,702
Natural Gas	48	100	252	613	1,157
Hydroelectricity	10	23	43	86	132
Total	1,485	1,793	2,610	4,197	6,306

Source: Joel Darmstadter et al., *Energy in the World Economy* (Baltimore: Johns Hopkins, 1971) Table 4.

Secondly, there were great variations between the main regions of the world, the older industrialized countries showing far lower rates of growth than the latecomers. Whereas the rate for Western Europe was 2.1 per cent and that for the United States 2.6, the Soviet Union's energy consumption rose at 9 per cent (with a sharp decline after 1950) and that of Japan at 5.3, with a very sharp rise after 1950. Figures for the Third World were also relatively high: Latin America 5.5, Africa 4.8, and non-Communist Asia 5 per cent.

Thirdly, the rates of increase of the various fuels differed very considerably. Coal and lignite grew slowly, at 1.5 per cent per annum over the whole period; after the war there was a significant rise in consumption, to 2.5 per cent for 1950-1965, but recently the rate of increase has dropped sharply, to 0.4 in 1965-68. Oil, on the other hand, has steadily accelerated, from 5.3 per cent in 1925-1950 to 7.6 in 1950-1965, and 8.5 in 1965-1970; for the period 1925-1970, the average was about 6.5 per cent.[4] As a result, whereas solid fuels contributed 83 per cent of world energy consumption in 1925, by 1950 their share had fallen to 61 per cent and by 1968 to 37 per cent. Conversely, that of liquid fuels rose from 13 to 28 and 43 per cent, respectively, and that of natural gas from 3 to 10 and 18 per cent.

Naturally, the fuel production of each country, or even of each region, has not grown in line with its consumption, and the balance has had to be met by the slowly increasing volume of world trade in coal and the very rapidly expanding trade in oil. Before the First World War, Western Europe (particularly Britain) and the United States were the main exporters of coal, while the United States and Russia accounted for most of the small amount of oil flowing into the world markets. The United States has maintained its position as a coal exporter and until 1949 was a net exporter of oil as well; however, in recent years its growing oil imports have turned it into an energy deficit area and in 1967 net imports accounted for 7.2 per cent of total United States energy consumption.[5] Western Europe had become a net energy importer by 1925 because of the growth of its oil consumption, and in 1967 over 60 per cent of its *total* energy needs were met by imports. In Japan the change was even more dramatic; by 1967 imports supplied over 80 per cent of energy needs. The Soviet Union, on the other hand, has succeeded in raising output of both coal and oil more rapidly than consumption and in 1967 exported nearly 12 per cent of its energy production. There are no other significant exporters of coal, and the energy deficits in both the advanced and the underdeveloped countries are met by a handful of oil exporters mainly in the Middle East, North Africa and the Caribbean. At present, imported fuels account for one-third of the world's total energy consumption. Correspondingly, fuels — mainly oil — have come to occupy an increasingly important position in international trade, their share in total value rising from under 5 per cent before the First World War and 6 per cent in 1929 to 10 per cent in 1967 and a distinctly higher figure in 1971.

At present, energy consumption per capita varies more widely than ever before. At one end, the United States is consuming well over 10,000 kilograms of coal equivalent, Western Europe 4,000 to 5,000 and the Soviet Union around 4,000. At the other, Latin America uses well below 1,000 and tropical Africa not much over 100 kilograms, with other underdeveloped regions falling in between. Both cross-sectional and times series regression analysis shows a high correlation between per capita income and energy consumption. For the world as a whole in 1950-1965, the Energy-GNP Elasticity Coefficient was 1.06, i.e., a one per cent increase in real gross national product was accompanied by a 1.06 per cent increase in energy consumption. But a regional breakdown shows significant divergences, the figures for North America and Western Europe being below 1, while those for the Soviet bloc, Latin America and non-Communist Asia were 1.2-1.3[6] In other words, high income countries consume less additional energy per dollar's worth of incremental income produced than low income countries. Similarly, historical studies for the United States, Britain, Germany and Belgium show that in the last fifty years or so their energy consumption per unit of income has been declining. This phenomenon is partly due to increasing "thermal efficiency" arising from technological progress and the substitution of more efficient for less efficient forms of energy. But an even more important factor has been greater "economic efficiency," i.e., the fact that a greater proportion of energy consumption is going to such uses as commercial space heating, where its efficiency is high, and a smaller proportion to mechanical work, where its efficiency is low. The latter trend is connected with the relative decline in the contribution of manufacturing and mining to gross national product and the rapid growth of services characteristic of mature economies; of course, manufacturing and mining consume more energy per dollar of product than do services. It should, however, be added that in the United States, during the last five or six years, the coefficient has once again been rising; this is mainly due to the spread of electricity, with its low efficiency, at the expense of other, more efficient but less convenient fuels.[7]

Coal, Oil and other Energy Resources

Coal was an almost indispensable precondition of industry for some two centuries; it is therefore no coincidence that all the major industrial powers (the United States, the Soviet Union, Germany, Britain, France, China, Canada and to a lesser extent Japan) have large reserves and an adequate production of coal.

The world's coal resources are enormous. At present, some 8,600 billion metric tons have been located, and it is estimated that as much will be found in unexplored areas, bringing the total to over 15,000 billion. These figures refer to coal in seams of a minimum thickness of 12 inches and at a maximum depth of up to 6,000 feet. But only about half the total is minable, and this is broken down as follows:

(billions of metric tons)

Soviet Union	4,310	Western Europe	377
Other Asia	681	Africa	109
United States	1,486	Oceania	59
Other North America	601	South and Central America	14

Calculations based on various assumptions show that both world and United States reserves are adequate to sustain, for another 200 years or more, an annual production of coal gradually rising to eight times the present levels.[8] But the fact that a mineral is physically recoverable does not mean that it can be produced economically, i.e., at a price high enough to cover its costs of production and transport yet sufficiently low to compete with available substitutes. Indeed, an oil industry source went as far as to claim that only 2.5 per cent of United States reserves "is really commercial production coal which you could produce and sell at a profit."[9] However, various extrapolations suggest that over the next few years production in Europe will decline but in the U.S. will rise appreciably, doubling by 1985.

Oil reserves are very differently distributed. Except for the United States, the Soviet Union and — very recently and to a smaller extent — Canada, no major industrial power has so far found large deposits or is a significant producer. Instead, the bulk of the world's known reserves are located in a relatively small number of underdeveloped countries, mainly in the Middle East, North Africa and the Caribbean (see Table 2). This dependence of the whole world, and more particularly its advanced regions, on a few, mainly small and vulnerable countries for an essential material of modern industry is one of the most characteristic and disturbing features of the oil industry. Estimates of the amount of oil that can ultimately be recovered are even more speculative than those for coal. The latest and most reliable, which were prepared by Standard Oil Company of New Jersey in 1967, are given below; the figure for the Middle East shows a sharp decrease over previous estimates, and may well be too low.[10]

(billions of U.S. barrels; 1 metric ton=about 7.3 barrels)

United States	200	Africa	250
Canada	95	Far East	200
Latin America	225	Soviet Union, Eastern	
Europe	20	Europe, China	500
Middle East	600	Total World	2,090

It should be added that other estimates differ significantly from these figures, the highest being about three times as large for both the United States and the world. But considering that in 1971 world output of oil was over 18 billion barrels (2.5 billion tons) and United States output 3.7 billion (532 million tons), it will be readily seen that present rates of production and growth can be sustained for only a few more decades.

Natural gas resources may be briefly mentioned. Although natural gas has been burned on a small scale in China and elsewhere for very long, until the Second World War transport difficulties and costs limited its use to a small radius around the field. But in the last twenty years large transcontinental pipelines have been built, first in the United States and more recently in the Soviet Union. The late 1960s witnessed the construction of several international gas pipelines, such as the ones between the Netherlands and other parts of Western Europe, between Iran and the Soviet Union, and between the Soviet Union and Eastern and Central Europe. In the last few years, technological improvements have made it possible to transport liquefied natural gas in tankers, some large contracts have been signed including one for supplying the United States with Algerian gas, and various projects have been discussed, including some for selling Soviet gas to the United States. This opens new perspectives for such areas as the Persian Gulf, whose gas output has been mainly flared up or reinjected into the fields for lack of large nearby markets. The world's proved reserves of gas are rising rapidly, but up to now seem distinctly smaller, in energy terms, than those of oil, let alone coal. A 1970 estimate put them at 45,670 billion cubic meters, equivalent to about 200 billion barrels of oil. Of these, 12,100 billion cubic meters were in the Soviet Union, 7,500 billion in the United states, 10,040 billion in the Middle East and 5,420 billion in North Africa.[11]

Lastly, there are fissionable materials. Presently established reserves of natural uranium are not large, and at foreseeable consumption rates would be exhausted in a few decades. Fortunately, developments in breeder reactors should expand enormously the supply of materials that can be used.

Ultimately, there is the distant promise of fusion, which, if fulfilled, could supply almost unlimited amounts of energy. In the meantime, atomic energy should increase very rapidly but is estimated to supply only about 35 per cent of electric power, and some 13 per cent of total energy, in the United States by 1985.

Projections to 1980

Projections of future consumption and production of energy in general, and oil in particular, are constantly being calculated and revised. The ones given below are the most recent available, but they are offered more as illustrations of probable magnitudes and trends than as exact forecasts of future needs and availabilities.

Several projections of total energy demand have been

TABLE 2

Estimates of Consumption of Petroleum Products in 1980

(millions of tons)

	Consumption in 1969	Estimated Annual growth rate (percent)			Estimated Consumption 1980		
		low	medium	high	low	medium	high
North America	764	2.0	3.9	5.2	950	1,164	1,335
Western Europe	547	3.1	7.4	10.3	765	1,200	1,608
Japan	160	7.9	12.7	16.6	369	596	867
Other Developed	44	3.6	6.8	10.1	65	91	127
Soviet Union	180[a]	....	8.2	....	....	550	....
Eastern Europe	39[a]	....	9.0	....	....	142	....
Developing Countries	203[b]	4.6	7.2	8.4	349	468	535
World[c]	(2,050)	....	....	....	3,269	4,317	5,285

a) 1965

b) 1968

c) Excluding "Centrally planned economies of Asia"; including bunkers

Source: United Nations, *Report of the Ad Hoc Panel of Experts on Projections of Demand and Supply of Crude Petroleum and Products* (ESA/RT/Meeting II/14), January 1972.

made for the major regions. First, as regards the United States, which in 1970 accounted for 31 per cent of world consumption, a compound growth rate of 4.2 per cent per annum has been projected for 1971-1985, giving a total increase of 84 per cent over the 1970 level.[12] For the European Economic Community, whose share of world consumption was 17 per cent, projected growth is higher, 5.6 per cent per annum in 1971-1985, for a total increase of 125 per cent.[13] The growth rates in other parts of Western Europe should be somewhat lower. For the Soviet bloc, the underdeveloped countries and Japan, growth rates are expected to be still higher, in ascending order.[14] For the world as a whole, the rate of growth may be about 5 per cent.

The most recent estimates regarding demand and supply for oil were made by a United Nations panel of experts and are shown in Table 2. Over the eleven-year period 1969-1980, they foresee increases ranging from a minimum of about 60 per cent, through a medium estimate of 110 per cent to a maximum of 160 per cent, implying annual compound rates of growth of about 4.5, 7 and 9 per cent respectively. These figures may be somewhat too high, since they do not seem to make enough allowance for the impact on oil markets of increasing consumption of coal and natural gas. As the table shows, the bulk of the increase will be absorbed by Western Europe and North America, followed by Japan and the Soviet Union. Forecasts of supply are even more tentative since they are liable to be affected not only by chance discoveries, technological changes and economic factors, but also by political upheavals. The estimates made by the same group of United Nations experts are given in Table 3. This brings out clearly the rapidly increasing importance of the Middle East and North Africa in oil production and the sharp decline in the western hemisphere's share (see also Table 5).

TABLE 3

Petroleum Production in 1960 and 1970 and Forecasts for 1980 and 1971-80, by Regions
(millions of tons)

	1960		1970		1980		1971-80	
	Output	Percent	Output	Percent	Output	Percent	Output	Percent
North America	390	37	605	26	850	19	7,335	21
Latin America	196	18	265	11	425	9	3,447	10
Western Hemisphere	**586**	**55**	**870**	**37**	**1,275**	**28**	**10,782**	**31**
North Africa	12	1	230	10	300	7	2,900	8
Rest of Africa	0.2	–	60	3	250	6	1,825	5
Middle East	262	25	690	30	1,790	40	12,187	35
Asia & Oceania	28	3	90	4	200	4	1,605	5
Western Europe	15	1	16	1	45	1	314	1
Soviet Union & Eastern Europe	162	15	374	16	640	14	5,112	15
World[a]	**1,065**	**100**	**2,330**	**100**	**4,500**	**100**	**34,725**	**100**

a) Excluding "Centrally planned economies of Asia"
Source: same as Table 2 above.

It also shows that in the coming decade the world will have to continue relying heavily on the Middle East and, to a lesser extent, North Africa. For the large increase in the western hemisphere's output (10,782 million tons, or 31.0 per cent of the world total) will be absorbed within its borders, and the same is true of the Soviet bloc (5,112 million tons or 14.7 per cent of the total).[15] This means that the needs of the rest of the world will have to be met by the Middle East, with 12,187 million tons or 35.1 per cent, North Africa (mainly Libya and Algeria) with 2,900 million tons or 8.4 per cent, the rest of Africa (mainly Nigeria) with 1,825 million tons or 5.3 per cent, and Asia and Oceania (mainly Indonesia) with 1,605 million tons or 4.6 per cent.

It goes without saying that such a huge increase in output of oil will require vast capital expenditures. Here forecasting is even more hazardous since the incremental investment per unit of capacity varies widely between regions, and also between the various activities such as production, refining, transport and marketing. An indication of the magnitudes involved is given by the fact that in 1969 total investment in the petroleum industry, excluding the Soviet Union, Eastern Europe and China, was put at $19.8 billion, of which $8.9 billion was in the United States. Other significant figures are the estimates of total investment required per daily ton of additional capacity (say, 2,500 barrels a year) for all phases from exploration to marketing; these varied between $15,000 and $28,000. The application of these ratios led the United Nations panel to suggest that new investment (i.e., excluding replacement) in the whole world in the 1970s might amount to some $170 billion, of which about one-eighth would be accounted for by the Soviet Union, Eastern Europe and China. Even higher figures for total gross investment were obtained by extrapolating the annual growth rate of 5.2 per cent in capital expenditures that prevailed in the 1960s. This suggests a total expenditure of some $270 billion in the 1970s excluding the Soviet Union, Eastern Europe and China. The magnitude of these sums reflects the vital importance of the petroleum industry in today's world.

The Oil Industry

From its inception a little over 100 years ago, the history of the oil industry has been one of uninterrupted growth. Production rose from 100 million tons in 1921 to over 250 million in 1938, and passed the 500 million mark in 1950, the 1,000 million in 1960 and the 2,000 million in 1969; it is now running at over 2,500 million tons.

The rapid progress made by oil is attributable to many causes. For some uses, such as lubrication and aviation or motor transport, oil has so far proved irreplaceable. For others, such as shunting locomotives, its efficiency is several times as great as that of coal. The fact that it is a fluid – and can therefore be made to flow through pipelines and into tankers and storage tanks – has helped to reduce its cost of transport per unit of energy content to about half that of coal, especially over long distances. Its fluidity also makes it easier to handle than coal, reducing labor costs in industrial uses and unpleasant work in domestic heating. Oil is a relatively clean fuel which leaves no ashes. Lastly, except in a few spots where large deposits of high-grade coal are available, oil is by far the cheaper fuel. These advantages have so far more than outweighed one of the main disadvantages of oil – the fact that it requires specialized carriers such as pipelines, tankers and tank cars, which necessitate a heavy capital outlay. Even this handicap is partly being surmounted by the use of tankers which can carry ores or grain as well as oil. Another disadvantage of oil – air pollution – is only just beginning to attract attention and remedial action, but it should be noted that oil is less polluting than coal and can be "cleaned" more easily.

Such rapid expansion of the output of a mineral presupposes the continuous discovery of new deposits, and in fact the main centers of production have shifted markedly. Until the early years of this century, the United States and Russia ran neck and neck for first place and between them accounted for the bulk of world production and exports. In the interwar period, the United States produced some two-thirds of the total, and important new sources were developed in the Caribbean and Middle East. After the Second World War, four centers — the United States, the Middle East, the Soviet Union and Venezuela — accounted for 85 to 90 per cent of output, but in recent years their share has fallen slightly because of the emergence of North and West Africa, Canada, Indonesia and other producers.

Today the petroleum world is divided into three main segments: the United States, the Soviet Union and the rest of the world. All have several common features: their oil industries are huge, very capital intensive, highly advanced and progressive in technology, and operate under conditions where competition is limited and output and imports are not allowed to vary freely and determine prices. The Soviet industry, which meets the needs of the Soviet Union and Eastern Europe, is of course state owned, and the volume of exports and imports does not seem to be determined by the relation between domestic costs and foreign prices. If it were, substantial amounts of Middle Eastern and North African oil might be imported since it would seem that it could be delivered to several parts of the bloc more cheaply than Soviet oil. In the United States, the large number of producers (over 10,000) makes for a certain degree of competition. But, since their collapse in the 1930s, petroleum prices have been sustained at an artificially high level by restrictions on both domestic output and imports. As regards output, the system of prorating practiced in almost all producing states has kept American production well below the level it would otherwise have attained, and also made it possible for tens or even hundreds of thousands of inefficient and high cost wells to continue operating. As for imports, in addition to a small duty levied since 1932, a quota system was introduced in 1955 on a voluntary basis, which was changed to a mandatory one in 1959. These quotas exempt Canadian and Mexican oil, give some preference to Venezuela, and restrict imports from other sources to a specified maximum percentage of United States production east of the Rockies. The result has been that in recent years the United States has imported over 20 per cent of its total consumption of oil, mainly from western hemisphere sources.

The outstanding characteristic of the third segment — which includes all the main exporters and importers — has been the predominance of the seven major international oil companies. Five of these are American: Standard Oil Company of New Jersey, Standard Oil Company of California, Socony-Vacuum Oil Company, Gulf Oil Corporation and the Texas Company. Two are British: Royal Dutch-Shell Oil Company and British Petroleum (formerly Anglo-Iranian Oil Company). An eighth, the French Compagnie Française des Pétroles, has generally cooperated with the others. Around 1950, these companies controlled nearly 90 per cent of oil production outside the United States and the Soviet bloc and almost 80 per cent of refining capacity. In addition, they owned or chartered two-thirds of the world tanker fleet.[16] They also controlled a very large proportion of the markets in the consuming countries. Since then, their share has somewhat declined, with consequences that will be examined later (see Chapter Three). But thanks to their control of the main sources of production and their anxiety to avoid price wars that would hurt them all, they managed to adjust output in the various regions in such a way as both to meet the rapidly rising world demand and to accommodate newcomers without disturbing the price structure.

The expansion of the oil industry has necessitated a huge amount of investment. The estimated value of gross fixed assets in petroleum (excluding the Soviet bloc) rose from $24.6 billion in 1947 to $97.3 billion in 1960 and $205.9 billion in 1970; the respective figures for net fixed assets were $12.2 billion, $53.2 billion and $113.7 billion.[17] About half the total is in the United States, one-sixth in other parts of the western hemisphere and one-third in the eastern hemisphere. In recent years, the bulk of investment in the western hemisphere, and well over half in the eastern, has been American, and has been mainly generated within the industry itself. The rate of return on such capital, although declining, still compares favorably with that on other foreign or domestic investment.[18]

CHAPTER TWO

Middle Eastern Oil: The Economics

The outstanding features of the oil industry of the Middle East are its huge reserves and very low costs of production; both these statements apply, to a lesser degree, to North Africa.

Reserves and Production

At the beginning of 1972, world proved reserves of petroleum were estimated at 632 billion barrels, or about 86 billion tons. Of these, 368 billion barrels, or 58 per cent, were in the Middle East and 42 billion, or 7 per cent, in North Africa. The remaining 35 per cent was divided between the United States, the Soviet Union, Venezuela, Indonesia and other countries (see Table 2). The region's reserves have risen rapidly in the last twenty-five years, both in absolute terms and as a percentage of world total. In 1936, Middle Eastern reserves had been put at under 5 billion barrels, or 20 per cent of the world total. By 1960 these figures had risen to 163 billion and 61 per cent, respectively, and another 8 billion barrels (3 per cent) had been discovered in North Africa.

It should be noted that the term "proved reserves" has a

TABLE 4

Estimated World Oil Reserves, 1 January 1972
(billion barrels)

United States	37.3		
Canada	8.5	Total	
Venezuela	13.9	Western Hemisphere	77.3
		Europe	14.2
		Total	
Indonesia	10.4	Asia-Pacific	15.6
Libya	25.0		
Algeria	12.3		
Egypt	4.0		
Nigeria	11.7	Total Africa	58.9
Soviet Union	75.0	Total	
China	20.0	Sino-Soviet	98.5
Saudi Arabia	145.3		
Kuwait	66.0		
Iran	55.5		
Iraq	36.0		
Neutral Zone	24.4		
Abu Dhabi	19.0		
Syria	7.3		
Qatar	6.0	Total	
Oman	5.2	Middle East	367.4
		Grand total, World	**631.9**

Source: *Oil and Gas Journal*, 27 December 1971.

special and rather restrictive meaning. It refers solely to those reserves that have been established by the exploring or producing companies or government agencies and that are judged to be economically recoverable with existing technology. Since the process of establishing reserves is expensive, and therefore not carried out much beyond anticipated needs, and since technological improvements are constantly increasing the proportion of oil that can be extracted from the various fields at a given price level, it is clear that the figures in Table 4 represent only a fraction of the amount of oil that can ultimately be discovered and recovered. It would therefore be fruitless to speculate on what the share of the Middle East and North Africa in proved reserves will be in, say, 1985. It is possible that discoveries elsewhere will outpace those within the region, and that the latter's share in the world total will decline (see Chapter One). But barring spectacular and utterly unforeseen discoveries of low cost oil, the region's predominance seems assured for at least another generation.

Two more facts deserve attention. First, Middle Eastern reserves, i.e., nearly 90 per cent of the regional subtotal, are located in the Persian Gulf area, and the oil produced from them can reach Western markets only through circuitous or vulnerable channels; North African oil, which accounts for just over 10 per cent of the subregional total, is however within easy reach of Europe. Second, except for the 56 billion barrels in Iran, all the region's oil, i.e., 86 per cent of the subtotal, is in Arab lands.

Output of oil in the Middle East and North Africa has risen, with a time-lag, along with its discovery. Commercial production started in both Egypt and Iran in 1908. Egyptian output remained low until the last few years, but Iran's increased rapidly after the First World War, and in the 1930s and 1940s it ranked as the world's fourth largest producer

TABLE 5

Production of Crude Petroleum
(thousands of barrels a day)

	1938	1949	1961	1971
Abu Dhabi	—	—	—	900
Bahrain	21	28	42	75
Dubai	—	—	—	126
Iran	210	560	1,201	4,514
Iraq	90	80	1,003	1,692
Kuwait	—	250	1,644	2,895
Neutral Zone	—	—	177	551
Oman	—	—	—	291
Qatar	—	2	178	429
Saudi Arabia	1	480	1,393	4,456
Syria	—	—	—	115
Total Middle East*	322	1,400	5,643	16,240
Algeria	—	—	327	603
Egypt	3	40	71	302
Libya	—	—	17	2,800
Total North Africa*	3	40	415	3,792
United States	3,480	5,480	8,174	9,650
Canada	20	60	643	1,336
Venezuela	520	1,320	2,923	3,579
Indonesia	160	120	517	880
Soviet Union	570	650	3,360	7,470
Total World*	5,595	9,750	23,547	47,800

Source: *Oil and Gas Journal; BP Statistical Review.*
*Includes other countries not listed.

after the United States, Venezuela and the Soviet Union. Iraq's oil began to flow abroad in 1934, but the amount produced remained small until the early 1950s, when the laying of large pipelines to the Mediterranean and the development of the southern oil fields made it possible to expand output severalfold. In the Arabian Peninsula, small amounts were produced in Bahrein and Saudi Arabia in the late 1930s, but it was only after the Second World War that output shot up in Saudi Arabia and Kuwait, in the 1950s in Qatar and in the 1960s in Abu Dhabi, Oman and elsewhere, including offshore areas. The 1960s also saw a very rapid expansion in Libya and a somewhat slower one in Algeria. Table 5 shows production in various countries in 1938, 1949, 1961 and 1971; it brings out the growing proportion of the Middle East and North Africa in world production (40 per cent in 1971), and the rising share of North Africa and declining share of Iran in the regional subtotal. In recent years, Middle Eastern production has been growing at over 10 per cent a year, and North African at over 20 per cent. Since the latter rate reflects both the initial spurt of development and the stimulus given by the closure of the Suez Canal in 1967, it clearly cannot be sustained and indeed is already declining. The Middle Eastern rate of growth, however, is probably close to the long-term trend for the next decade or two. Since the corresponding rate for world production is put at about 7 per cent, the region's share should increase steadily and appreciably (see Table 3).

Costs of Production

Estimates of costs of production differ significantly since they are based on widely different assumptions and methods, but all agree in putting the Middle Eastern figures at only a small fraction of those in other regions. Perhaps the most thorough estimates are those of Paul Bradley, who put average development costs per barrel in 1953-1962 at $.11 to $.15 for the major Middle Eastern producing countries and $.22 for Libya, compared with $.39 for Venezuela. Other estimates, by M.A. Adelman, give even lower figures for 1962-64: $.04 to $.10 for the Middle East, $.15 for Libya and $.46 for Algeria, as against $.62 for Venezuela and $1.51 for the United States.[1] The comparable figure for the Soviet Union is about $.80. Both economists agree with others that the Middle Eastern marginal cost curve will not rise appreciably and may indeed decline if production is expanded substantially; in other words, the Middle East should be able to supply a far larger volume of oil at a cost not much above the present one, or even somewhat below it.

The main reason for the low costs of production in the Middle East, and to a lesser extent in Libya, is the very high yield per well. This in turn is attributable to two sets of factors, natural and institutional; lower labor costs are an insignificant factor since labor constitutes a tiny fraction of total costs (about 5 per cent) in such a capital-intensive industry as oil, and since wages in the Middle Eastern and North African oil industries are relatively high. (The average annual compensation per Saudi employee at Aramco, in Saudi Arabia, is now over $3,500.)

As regards natural conditions, with the exception of the northern Iraqi fields, almost all the Middle Eastern oil deposits that are under exploitation are close to the coast, which means an economy in pipeline and pumping costs. Most of them are not too far beneath the surface and have not necessitated very deep drilling — generally about 5,000 to 6,000 feet. Above all, the bulk of the oil lies in the world's largest discovered pools. In 1969, it was estimated that of 71 fields in the world outside the Sino-Soviet bloc with over 1 billion barrels of oil, 38 were in the Middle East and 7 in

North Africa. The largest, the Burgan field in Kuwait, was estimated to hold 62 billion barrels, Ghawar in Saudi Arabia 45 billion, Safaniya, also in Saudi Arabia, 25 billion and Kirkuk and Rumaila in Iraq 15 and 14 billion respectively. By contrast, the biggest field yet developed in the United States, East Texas, is estimated to have originally contained only 6 billion.[2] Lastly, the porosity of the rock formations and the very great gas pressure make it possible to extract oil far more cheaply than in most other parts of the world.

As regards the institutional factor, by a fortuitous set of circumstances the Middle Eastern oil industry was established and operated under conditions that — with all their financial, social and political shortcomings — assured the maximum amount of technical and economic efficiency. The same applies to a lesser extent to Libya, and until recently, also held true for Algeria. It arose because of the nature of the concessions granted by the Middle Eastern and North African governments.

Taking the world as a whole, the oil industry operates four legal systems — Anglo-Saxon common law, Soviet, civil, and Muslim — which have resulted in three different patterns of ownership and concessions. In the United States, ownership of the subsoil and the minerals it contains goes with the ownership of land. Hence the oil deposits belong to many thousands of landowners each of whom makes his own arrangements with the persons or companies proposing to drill for oil; generally, each receives a fixed royalty in cash or kind. This has made possible the entry into the oil industry of numerous small operators, in addition to the large companies. (At the present time there are over 10,000 small operators.) But it has also led to the sinking of millions of wells since each landowner is interested in maximizing the output from his own property and, if possible, in drawing away some of the oil lying under his neighbors' land — the so-called offset wells. At present, some 517,000 wells are in operation in the United States, with an average production of only 19 barrels a day. Three major disadvantages follow: the high capital investment represented by the multitude of unnecessary wells; the high cost of production per barrel of oil produced, since the total cost of operating a well with a very large output is only a little above that of a well with a very small output; and the loss of gas pressure, and hence of recoverable oil, due to the presence of these wells and to the failure to treat each oil field — rather than each plot of land — as a producing unit.[3]

Under Soviet law, on the other hand, minerals belong to the state, and the Oil Trust is, presumably, free to practice unitization of oil fields and seek to produce at the maximum efficient rate (MER). But, like other industries, petroleum has suffered from certain micro-economic inefficiencies inherent in the Soviet system, such as irrational pricing and "success indicators" that induce economically irrational responses. Thus, to take one instance, "plan fulfillment in prospecting enterprises is chiefly measured by the number of metres drilled — with the result that drillers who run into difficulties often leave a well unfinished and start a new one in order to achieve a large number of metres."[4] Such handicaps, and the government's mistaken decision in the 1930s to meet the country's energy needs from coal rather than oil, retarded the Soviet petroleum industry's development until the last few years.

In other producing areas, where civil or Muslim law prevails, the subsoil and its minerals also belong to the state, but actual operations are carried out by private enterprise. This has meant that their oil industry was developed under a system of concessions granted by their governments to foreign oil companies. Moreover, the political relations and ideologies prevailing in the Middle East at the time led to concessions covering huge areas of land — up to several hundreds of thousands of square miles — and for durations of sixty to ninety years. Until recently, all the concessionary companies belonged to the Big Seven American or British companies or their affiliates, with which the Compagnie Française des Pétroles worked closely. This made it possible to combine the technical advantages of both the American and Soviet systems. On one hand, the industry benefitted from the resources, flexibility and advanced technology provided by one of the most dynamic of modern capitalist industries. On the other hand, these companies, enjoying a long time horizon and untrammeled by individual property rights and other legal impediments, could unitize production and sink only the number of wells required for optimum operation. Again, an example may be given for illustration. Although they are different in certain respects, Abqaiq field in Saudi Arabia and East Texas are approximately equal in surface area (though Abqaiq has distinctly larger reserves); yet, even after the improvements mentioned earlier, in 1952 East Texas had no less than 26,000 producing wells, compared to 62 at Abqaiq.[5]

Three important consequences follow from this. First, there has been a great saving in capital costs due to the much smaller number of wells drilled; thus it was calculated that in 1959 gross fixed assets per barrel of daily crude oil capacity amounted to only $290 in the Middle East, compared with $1,340 in Venezuela and $3,190 in the United States.[6] Secondly, due in part to the favorable geological conditions mentioned above, output per well was much greater — about 6,500 barrels a day in the Middle East in 1971 compared to under 400 in Venezuela and 19 in the United States. Moreover, almost all the oil produced in the Middle East has come from "free-flowing" wells, i.e., the oil is forced up by gas pressure; by contrast, some nine-tenths of United States wells, three-fifths of Venezuelan and a substantial proportion of Soviet wells require pumping, which naturally adds to costs. Lastly, it may be presumed that unitization will result in the ultimate recovery of a larger proportion of the "oil in place" than in the United States.

The situation in Libya and Algeria is less favorable than that in the Middle East but distinctly better than in other regions. For one thing, the fields are smaller: Sarir, Zelten and Gialo in Libya have estimated reserves of 8 billion, 2.2 billion and 2 billion barrels respectively, and Hassi Messoud in Algeria, 2.7 billion. The relatively large number of concessions granted to foreign companies — which has had great financial advantages for the governments — and the desire of some companies to secure maximum possible output may have led to less than optimal operations; at any rate, the number of producing wells is relatively greater and daily output per well — 2,800 in Libya and 1,200 in Algeria in 1971 — is distinctly lower. Lastly, about one-third of the oil produced in Libya is pumped, but in Algeria the bulk comes from free-flowing wells.

Costs of Transport

Since oil is a bulky commodity, transport charges form a large part of delivered costs. It is clear that for the Japanese market — the second largest outside the Soviet bloc and the most rapidly growing — as well as for the Indian Ocean region, the Middle East is better situated than any exporting area except Indonesia. On the other hand, it is less well placed than either North Africa or Venezuela as regards Western Europe and the eastern United States, and the closure of the Suez Canal has further increased its disadvantage.

It is difficult, however, to give exact and meaningful

figures because of the nature of the tanker market. The world tanker fleet — 156 million tons in 1971 — falls into three categories: oil-company-owned tonnage, accounting for 35-40 per cent; medium and long-term charters, representing 40-45 per cent; and tonnage chartered on a short-term (up to two years) or spot basis, which fluctuates between 5 and 20 per cent of the total. The factors determining freight rates differ sharply in the three categories. Transport costs in company-owned ships are largely determined by building costs and operating expenses. In normal times, medium and long-term charter rates also tend to approximate building and operating costs and to be fairly steady. But short-term and spot rates fluctuate violently, since the impact of a sudden increase or decrease in demand for tankers is felt initially in that segment of the market.[7]

One further factor should be noted: real costs of transport have been falling considerably and, except for interruptions caused by such events as the two closures of the Suez Canal, the stoppage of the Syrian pipelines or other political disturbances, the trend of tanker freights has been downward. This has been partly due to the great reduction in building costs (e.g., the cost per ton of a 45,000 dwt tanker fell from $240 in 1956 to $110 in 1963) and partly to the huge economies of scale derived from the great rise in the average size of tankers. Whereas in the early 1950s the typical tanker had a tonnage of 10,000 to 25,000 dwt, by 1970 such ships represented less than 20 per cent of the total tonnage, while those of 200,000 or more accounted for over one-fifth of existing tonnage and for nearly 90 per cent of shipping under construction. In 1969 the building cost of such tankers was put at only $75 per dwt, and their operating costs are also relatively very much lower than those of smaller ships.[8]

The consequent decline in freight rates may be illustrated by the fall in time charter rates per ton from the Persian Gulf to the east coast of the United States, from $1.25 to 1.30 a barrel in 1952 to $.25 to $.60 in 1966. This naturally bene-fitted the Middle East and enabled it to extend its range of competitiveness. In 1966 it was estimated that, as regards Western Europe, Libya's freight advantage over the Middle East was $.37 a barrel and Venezuela's advantage $.23; in other words, the Middle East's lower production costs more than offset its freight disadvantage compared to Venezuela but not to Libya. The closure of the Suez Canal in 1967 raised these figures to $.54 and $.38 respectively,[9] but the gap has been somewhat narrowed since by the increasing size of the tankers carrying Persian Gulf oil around the Cape. The reopening of the canal and the completion of the plan to enlarge and deepen it to accommodate tankers of 260,000 dwt loaded and 290,000 in ballast — originally scheduled for 1976 — would of course greatly reduce the cost of trans-porting Middle Eastern oil to the European, Soviet and American markets, as would the projected trans-Egyptian pipeline.[10]

Prices

This is not the place to go into the complex question of oil pricing; all that need be noted here is that various forces have kept the price of Middle Eastern and North African oil far above its costs of production, though well below that of competing fuels. First, the establishment of prorationing in the petroleum industry in the United States in the 1930s restricted supply and raised the price of oil in this country. Since the United States was by far the leading exporter of oil, United States prices had been taken as a base for fixing the price of Persian Gulf refined products in the 1920s. When, during and immediately after the Second World War, the Middle East became a large exporter of crude oil, the posted price of the latter was also based on Gulf of Mexico prices. After 1948, however, the two series began to diverge, United States crude prices rising steadily while Middle Eastern tended to decline, but they still showed sympathetic short-term fluctuations, as in 1953, 1957, and 1959. The decline in its price — from $2.04 a barrel for Iranian light crude in 1948 to $1.78 in 1961 — enabled Middle Eastern crude oil gradually to capture practically all the markets of the eastern hemisphere, leaving to Venezuela only the United States, Canadian and part of the Latin American markets. But Middle Eastern oil remains shut out of two important potential markets: the Soviet, because of the autarkic policy of that government, and the United States, because of the import quotas (see Chapter One). Yet there is no doubt that Middle Eastern oil could successfully compete in both. Thus in 1969 it was estimated that Middle Eastern heavy crude could have been delivered, by very large tankers, to the United States east coast at as little as $1.60-1.70 a barrel and Libyan at $1.70-1.80, compared to about $3.00 for Texas oil at the same point of delivery.[11]

There is good reason to believe that, in the rest of the world, the price elasticity of demand for crude oil is low and that therefore a reduction in price would not have been more than compensated by greater sales. Hence there was every incentive to keep the price of Middle Eastern oil high. And the fact that the bulk of production and marketing, both within the Middle East and in the rest of the world outside the United States and Soviet bloc, was controlled by the Big Seven oil companies, meant that output and sales could be adjusted so as not to lead to sudden disruptions of supply and demand and to sharp changes in prices. By the late 1950s however, increasing pressure began to be felt because of greater output and sales by independent producers, rising exports by the Soviet Union and growing competition between the large oil companies. The development of Libyan oil added to the pressure. Hence, Middle Eastern posted prices were repeatedly reduced and, in addition, increasing discounts were granted to an ever widening circle of buyers. By 1967, although the posted price of Iranian light crude (34°) was $1.79, discounts were estimated at 45 to 50 cents, giving an effective price of $1.29 to $1.34.[12]

This pressure, the still very wide gap between costs and effective prices, and the prospect of increasing competition in the industry and a consequently greater pressure of supply, led several observers − notably M.A. Adelman − to predict that prices would continue to decline appreciably.[13] But this was to ignore the political factors at work, and the situation was abruptly reversed by the events of late 1970 and 1971 (see below). The result was a sharp rise in posted prices: the price of Iranian light crude had, by June 1971, advanced to $2.27 and other Persian Gulf crudes showed a corresponding rise of some 30 per cent. North African crude went up even more, Libyan light oil (40°) rising by some 55 per cent, from $2.21 in June 1970 to $2.55 in January 1971 and $3.42 in June 1971. The drop in the value of the dollar in the fall of 1971 led to a further increase in posted prices of 8.5 per cent in January 1972.[14] But it should be noted that some discounts are still being granted to buyers.

Concessions

As mentioned before, the concession agreements granted to the oil companies in the Middle East and North Africa gave the producing countries the maximum efficiency of operation. But they were far from satisfactory either politically or financially. As regards politics, the presence of these huge international giants, generally having more political and economic power than the government of the producing country, was inevitably felt as a threat and irritant by local nationalists. This has been a worldwide phenomenon with counterparts in such countries as Rumania, Mexico, Venezuela, Indonesia and Peru.

Nor, until the last twenty years, were the financial arrangements more satisfactory. The concessions under which the industry operated until 1950 had been granted at various times between 1901 and 1935, but they had one feature in common: they stipulated payment by the companies to the host governments of a fixed royalty of around 4 gold shillings a ton, i.e., 20-25 cents a barrel.[15] On average, the Middle Eastern governments received 21 cents a barrel, a figure close to the Venezuelan one of 23 cents.[16] In return, the companies were exempted from all taxes, including customs duties, and were granted complete power in determining production and pricing policy. These arrangements may have been fair enough at a time when demand was growing slowly and prices were fluctuating or sagging, when the prospects of striking oil seemed very uncertain, and when huge capital investments had to be made. But the sharp rise in prices, including that of oil, during and after the Second World War reduced both the purchasing power of the fixed royalty and the governments' share of the value of each barrel of oil – in Iran, for example, according to a government estimate, from one-eighth before the war to one-sixteenth in 1947.

The growing discontent first came to a head in Venezuela in 1943, when taxes were imposed which eventually gave the government half the net income earned by the industry, thus raising payments per barrel to 70-80 cents. In the Middle East, a similar arrangement was concluded with Saudi Arabia at the end of 1950; still earlier, in 1948 and 1949, concessions had been granted in the neutral zone lying between Saudi Arabia and Kuwait which offered royalties more than twice the current level. In the meantime, negotiations had been initiated in Iran for revision of the existing agreement, and the volatility and explosiveness of Persian politics together with the rigidity of the company led to the crisis of 1951, when the industry was nationalized and production brought to a standstill. This crisis alerted the other companies, and by the end of 1952 the so-called fifty-fifty arrangement had been extended to all producing countries and was also applied to Iran when the dispute was finally settled in 1954.

The result was to raise payments per barrel to 70-80 cents, and since output was also increasing very rapidly, the total revenues accruing to the Middle Eastern governments multiplied almost tenfold between 1948 and 1960 — from about $150 million to nearly $1,400 million. The companies' share was almost halved, falling from an average of about $1.52 per barrel in 1948 to about $.87 in 1960. But, thanks to the very low costs of production and the great expansion of output, company profits remained very high. It has been estimated that in 1948-1960 the companies' net income from Middle Eastern oil operations averaged 60 per cent of their net assets in the region.[17] Even granting that this calculation does not take account of the considerable assets outside the region required for the carrying, refining and marketing of Middle Eastern oil, and on which profits were much lower and occasionally negative, it does indicate the very high level of the companies' earnings in those years. And it should moreover be remembered that the companies were allowed to offset their payments to the Middle Eastern governments against their tax liabilities in the United States and United Kingdom, thus further enhancing the profitability of their Middle Eastern operations.

The acceptance of the profit-sharing principle introduced a new element into the Middle Eastern picture: it made the governments keenly interested in the pricing of oil. Hitherto their revenues had varied solely with volume of output; now revenues depended on price as well, since the sum that was split was the difference between the posted selling price of oil and its cost of production. Hence the price reductions of 1959 and 1960 led them to seek a way of influencing prices, and in September 1960 Venezuela and a group of Middle Eastern countries formed the Organization of Petroleum Exporting Countries. By now OPEC includes not only all the major Middle Eastern and North African oil producing countries but Indonesia, Nigeria and Venezuela as well and accounts for over half the world's output and about 90 per cent of exports. In its first ten years OPEC failed to achieve its announced objective of restoring prices to their 1959 level, though it did stop further declines in posted prices. It also was unable to implement its 1965 resolution (Resolution IX.61) calling on members to adopt "a production plan calling for rational increases in production from the OPEC area to meet estimated increases in world demand," since the diverging interests of its members — arising from great differences in size of reserves, costs of production, access to certain markets and need for oil revenues — made them reluctant to agree on a common production plan. But it did succeed in gaining important financial concessions, notably the companies' consent to the "expensing of royalties," i.e., that the 12.5 per cent royalty payment which the governments were entitled to receive in cash or kind were not to be treated as a credit against the companies' liabilities to the government as part of the latter's 50 per cent share, but as costs of production to be paid to the governments in addition to the 50 per cent share.[18] Thanks to these gains, Middle Eastern government receipts per barrel of oil exported rose from an average of 75.7 cents in 1961-62 to 84.6 cents in 1968-69, and in Libya from 63.7 to 100.4 cents,[19] in spite of the downward trend of "effective," as distinct from posted, oil prices (see above). Consequently, the governments' share of profits rose from the 50 per cent level originally envisaged in the agreements to about 70 per cent.

Mention may also be made of the Organization of Arab Petroleum Exporting Countries. OAPEC was founded in 1968 by Saudi Arabia, Kuwait and Libya; in 1970 it took in Algeria, Abu Dhabi, Bahrain, Dubai and Qatar and in 1971 amended its charter to make possible the admission of all other Arab oil exporting countries. It differs from OPEC in being restricted to Arab countries but also in being empowered to establish companies directly concerned with oil operations. So far, it has considered proposals for joint pipelines, a tanker fleet and chemical fertilizer factories, but none of these has as yet been implemented.

In the meantime, two further significant developments were taking place: the creation of national oil companies and the emergence of new types of concession agreements. As regards the first, the creation of the National Iranian Oil Company in 1951 was followed by similar ones in Kuwait in 1960, Iraq in 1961, Saudi Arabia in 1962 and Libya in 1968. In Algeria, Sonatrach, the national company founded in 1963, played a leading role until 1971 when the nationalization measures taken by the government gave it a controlling interest in all oil activities in the country. These companies undertake a dual function: they represent their governments in the partnership and contract agreements discussed below, and at the same time, carry out various oil operations directly.[20] As regards the latter, some of the companies have engaged in production, refining, transport and marketing within their own countries; some have promoted petro-chemical and other industries; some have acquired the nucleus of a tanker fleet; some have participated in building refineries in consuming countries, e.g., in India, South Africa and Rhodesia; and some have entered into barter agreements with Eastern European and other countries. One of the most important aspects of their activities is that they have trained hundreds of experts in various branches of oil operations and have involved their governments in the world oil market; to that extent they have prepared the ground for a smoother eventual take-over of the oil industry by the various Middle Eastern and North African governments.

Iran also pioneered the partnership or participation form of agreements in its contract with the Italian AGIP Mineraria in 1957. At present over twenty such concessions have been given in the Middle East and Libya, almost all to relatively small companies of various nationalities – American, French, British, Italian, Japanese, German, Indian and Spanish.[21] Although the details vary considerably, certain common features stand out. First, these concessions cover offshore areas – or else areas not granted to the original concessionnaire or relinquished by it. Secondly, the areas included are relatively small and the duration of the concession is relatively short, usually twenty-five years after oil discovery. Thirdly, the companies bear all exploration costs, and often pay substantial bonuses as well; in addition some of them have offered "sweeteners," such as help in setting up petrochemical industries. Fourthly, the governments have the option of acquiring a participatory interest ranging between 20 and 50 per cent, usually payable in oil. Fifthly, these agreements show great flexibility regarding tne price to be taken for calculating receipts and profits; usually the company is allowed to use "effective" rather than posted prices. Lastly, under these agreements, the governments have a twofold source of revenue: the usual per cent share of the company's profits and the amount accruing to the government on its own equity share. Unless the discounts allowed by the company are very great, the sum of these two shares should substantially exceed the amount accruing to the government under the older "fifty-fifty" agreements.

Starting in 1966, a few contract agreements have also been signed by Iran and Iraq, mostly with French companies. Under these, the company is a contractor, not a partner, but here too it bears the initial exploratory costs. If oil is discovered, the company will advance to the host government the sums required for development, charging interest. In return, it will be entitled to buy, at cost plus taxes, a specified share of the oil extracted. So far, no oil has been produced by any contracting company so it is impossible to say how these terms compare with the older concessionary agreements. Most recently, an agreement was signed between the Soviet Union and Libya for developing the latter's fields and refineries, but the specific terms have so far not been revealed.[22]

The slow process by which the Middle Eastern governments were acquiring an ever increasing share of the economic rent produced by the low Middle Eastern costs of production was greatly accelerated in 1970. The closure of the Suez Canal and the shortage of shipping had caused spot freights to shoot up to nearly six times their 1967 level. The producing companies responded by raising output in Libya from 1,500,000 barrels a day in 1966 to over 3,100,000 in 1969, filling about 30 per cent of Europe's requirements from that source. This rendered them dangerously dependent on Libya, a dependence further aggravated by two factors: an unexpectedly large rise in world demand for oil and the diminution of two other sources of supply — Nigeria, because of the civil war, and Saudi Arabia because of the "accidental" damaging of the pipeline in Syria. The revolutionary government of Libya, which had come to power in 1969, quickly seized the opportunity offered. First, in the name of conservation, it forced the companies to reduce output and then it put pressure on the one most dependent on Libyan production and therefore most vulnerable, Occidental Petroleum. The other companies soon had to follow suit, accepting the raising of the income tax rate to 54-58 per cent and that of posted prices by 30 cents or more. Concurrently, the government of Algeria was demanding that the French companies operating in that country increase their "tax reference" prices, i.e., the ones used as a basis for calculating profits.

These developments made it possible for OPEC, in its Tehran meeting of January 1971, to register its greatest success to date. The Persian Gulf countries obtained an increase in their share of profits to 55 per cent and a rise in posted prices of 35-40 cents, to be raised by another 5 per cent a year until 1975. In return, the governments undertook not to make any new demands until 1975 but, as noted earlier, in January 1972 they obtained an increase of 8.5 per cent to offset the devaluation of the dollar. A rough calculation suggests that at present the Middle Eastern governments are getting $1.40 per barrel of oil produced. Meanwhile, in April 1971 Libya had obtained a large increase in posted prices.

Impact on the Region

The great expansion of oil production and the increasingly favorable terms obtained by the governments have led to a huge rise in the revenues received by them. Table 6 gives the available data and shows that the revenues of the Middle Eastern governments increased from $1,500 million in 1961 to nearly $4,200 million in 1970. The latter figure does not reflect the effects of the Tehran agreement, and it is probable that by 1975 the Middle Eastern total will be over $10 billion. Libyan revenues should also increase in about the same proportion, and Algerian oil and gas revenues, which

TABLE 6

Government Oil Revenues
($ million)

Year	Kuwait	Saudi Arabia	Iran	Iraq	Abu Dhabi	Qatar	Others*	Total ME	Libya
1961	464	400	301	266	—	53	13	1,498	3
1962	526	451	334	267	3	56	13	1,649	39
1963	557	502	398	325	6	60	13	1,861	109
1964	655	561	470	353	12	66	14	2,131	197
1965	671	655	522	375	33	69	16	2,342	371
1966	707	777	593	394	100	92	19	2,682	476
1967	718	852	737	361	105	102	24	2,898	631
1968	766	966	817	476	153	110	83	3,370	952
1969	812	1,008	938	484	191	115	118	3,666	1,132
1970	897	1,200	1,076	513	231	122	150	4,189	1,295

*Bahrain, Oman (beginning in 1967) and Dubai (beginning in 1969).

Source: *Petroleum Press Service*, September 1971.

amounted to $580 million in 1970, are estimated to reach $1,000 million in 1972.[23] This enormous flow of funds has, naturally, transformed the economies and societies of the producing countries. Perhaps the most dramatic example is that of Libya, whose per capita income rose from about $50 in the 1950s to $500 in 1965 and over $1,000 in 1968, and is now approximating $2,000. Concurrently, school population rose from 33,000 in 1951 to over 250,000 in 1966, and has continued to grow. But similar changes can be observed in all the oil countries.

The extent to which the oil industry has become the mainstay of these economies, and the consequent dependence of the countries concerned, has been thoroughly studied by Schurr and Homan, whose results may be briefly summarized.[24] By the late 1960s, oil contributed some 75 per cent of the gross foreign exchange earnings of Iran and Iraq, between 85 and 90 per cent in Kuwait, Libya and Saudi Arabia and close to 100 per cent in the smaller sheikhdoms. Oil revenues accounted for over 60 per cent of government revenues in Iran and Iraq, 85 to 90 percent in Kuwait, Libya and Saudi Arabia, and close to 100 per cent in the sheikhdoms. And although the industry is very capital intensive and employs relatively few workers, total employment by the foreign oil companies in these five countries in 1967 was 60,000 — incidentally showing a steady decline from 77,000 in 1961.

Schurr and Homan's figures may be supplemented by a few others. In the late 1960s, the oil sector accounted for just under 20 per cent of GNP in Algeria and Iran, 33 in Iraq, 55 in Saudi Arabia, 60 in Libya, and much higher ratios in Kuwait and the other sheikhdoms. Towards the end of 1971 the gross foreign exchange holdings of these countries were, in round numbers: Algeria $400 million, Iran $400 million, Iraq $500 million, Kuwait $1,500 million (plus over $1,000 million in government assets), Libya $2,600 million and Saudi Arabia $1,300 million.[25] It goes without saying that the new agreements, and the anticipated increase in output, will greatly raise all these amounts and percentages, except those pertaining to employment.

A few words may be added on the impact of oil on the other countries of the region. In the late 1960s the transit countries received significant revenues from pipeline tolls — Syria some $60 million, Lebanon $15 million and Jordan $5 million[26] — and these sums are due to rise under recent agreements. Similarly, until the closure of the Suez Canal in 1967, Egypt was earning over $200 million a year in dues levied on tankers. The large pipeline that carries oil from Elat to the Mediterranean brings a certain amount of revenue to Israel, and the one being constructed from Suez to Alexandria provides for a payment of $1.45 per ton to the Egyptian government. Egypt, Syria and Tunisia are exporting relatively small but increasing amounts of oil, which are beginning to make a significant contribution to their balance of payments. Further indirect benefits to all these countries include: exports (running in the tens of millions of dollars a year) of foodstuffs and industrial goods to the oil producing countries; the remittances sent by their nationals working in the Persian Gulf and Libya (which also add up to tens of millions and are especially important for Jordan); and the vast sums spent by nationals of the oil producing countries in them for tourist or other services or for the purchase of real estate or other investments, a factor of great significance for Lebanon. Lastly, the Kuwait Development Fund has advanced to the other Arab countries a total of $240 million, and the Kuwaiti, Libyan and Saudi Arabian governments have granted others — notably the Egyptian and Jordanian — sums running into several hundreds of millions. Altogether, directly and indirectly, the income generated by the oil industry in the region has become the mainstay of its economy, and its elimination or curtailment would constitute a crippling blow.

CHAPTER THREE

Middle Eastern Oil: The Politics

A n analysis of the political forces impinging on the Middle Eastern and North African oil industry can begin with a scenario of the likely interplay of those endogenous to the industry, i.e., between governments, companies and consumers. Such a scenario can both clarify the analysis and present what is probably the most optimistic picture than can be hoped for.

Endogenous Forces

As early as 1960, it had become clear to some observers that "the conditions under which bilateral monopoly bargaining is being carried out between companies and governments are changing in favor of the latter" and that "an increasingly large share of the economic rent derived from the lowness of Middle Eastern production costs and from the price structure will be absorbed by the governments."[1] By the mid-60s it was plain that: "the twin pillars which held the conduit — Anglo-American military and political power in the Middle East and North Africa and control of production by the American and British companies — are being undermined."[2] Three aspects of this process may be distinguished. First, there is the change in the balance of internal forces within the region; secondly, there is the great weakening of the Anglo-Saxon powers, which formerly dominated its politics and still control its oil. Both these sets of forces will be discussed below. Lastly, there is the growing competition in the world oil industry outside the United States and Soviet bloc and the declining part played by the major companies. In production their share fell from 82 per cent in 1963 to 77 in 1969, that of the independents rising from 9 to 14 and that of governments from 9 to 10; in refining the drop was from 65 to 56 per cent and in marketing from 63 to 54.[3]

By now, however, it seems evident that the governments of the producing countries have a second aim in addition to securing larger shares of profits: the absorbing of some of what may be called the consumers' rent enjoyed by the oil importing countries because of the cheapness of oil compared to other fuels; it seems clear that, if forced to, consumers would pay much higher prices for their essential oil supplies. This has been partly achieved by the 1971 price rises. It has been calculated that, for Western Europe, "by 1975 the increased cost of the new OPEC settlements will be as much as $5.5 billion [over the 1970 figure of about $9.5 billion]; for Japan the increase will be over $1.5 billion (the oil import bill was on the order of $2.5 billion last year). For the developing countries whose oil import costs were around $2.1 billion in 1970, the cost of the OPEC advances will be about $1 billion for 1975."[4] Since then prices have been raised by an additional 8.5 per cent in dollar terms (see Chapter Two) and one can presume that further rises will take place after 1975, or perhaps before. The companies are not in too good a position to resist such increases, and their incentive to do so is reduced by the fact that they can pass them on to consumers, who will have to take them, while, of course, making every effort to cut their oil consumption and shift to other fuels.

Nor are the companies in a much better position to resist an increase in the rate of taxation of profits beyond the present 55 per cent level. It is true that earnings of the seven international oil companies per barrel of oil produced in the eastern hemisphere fell from 56.5 cents in 1960 to 33 in 1970 (though preliminary indications suggest a rise in 1971). However, the huge increase in output raised their total earnings on such production from $1,101 million to $1,882 million.[5] Taking the Middle Eastern oil industry as a whole, net earnings on production in 1970 may have been about $1,650 million, which, on a net investment of about $4,000 million, suggests a rate of return of 40 per cent − well below the figures for the 1950s and early 1960s but still substantial.[6]

In the meantime the governments are advancing a further claim − participation in the ownership of the companies. This idea began to be forcefully expressed early in 1968, its main spokesman being Ahmad Zaki Yamani, the oil minister of Saudi Arabia. In June 1968, OPEC supported the principle in a Declaratory Statement. And in September 1971, at its Beirut Conference, OPEC passed the following resolution (XXV.139): "that all Member Countries concerned shall establish negotiations with the oil companies, either individually or in groups, with a view to achieving effective participation on the bases proposed by the said Ministerial Committee."

Participation has an obvious appeal to the governments.[7] It improves their image at home and in neighboring countries by showing that they can successfully oppose foreign control. It provides a means for operating in the world oil market and gaining valuable experience. It promises more effective control of the activities of the companies, including pricing and investment policies. Lastly, it should increase government revenues. But participation also raises many knotty questions. First, there is the desired participation percentage: the most commonly quoted figure is 20 per cent, but Algeria has already obtained 51 per cent, Libya is aiming at the same figure, and Nigeria is reported to be demanding an initial 35, to be eventually raised to 51.[8] Clearly a government minority interest presents fewer complications to the oil companies and is more readily acceptable to them than a majority interest. On the other hand, they may simply regard it as the thin end of the wedge − and indeed it has been indicated as such by some government spokesmen. But, when all is said, all signs point to an acceptance by the companies of 20 per cent in the n ar furture.

Secondly, there is the scope of participation. At one time, spokesmen of the producing countries were urging participation in "downstream" operations − i.e., transport, refining and marketing − as well as in production; but recently they seem to have given up this idea. This is due in part to realization that larger profits are made at the production stage. In addition, as pointed out by Nadim Pachachi, secretary general of OAPEC, "Why should we invest our money in countries like Western Europe or Japan when the Arab countries feel the need to invest them at home and develop refining and petrochemical industries on their own territories?"[9] Furthermore, the governments can see that downstream participation would render them more vulnerable. As *The Economist* (3 July 1971) pointed out, "They might be less inclined in future to expropriate us if we could in return expropriate them."

A third problem with participation is the basis on which compensation for the equity share would be made to the companies. Several possibilities have been suggested but none is satisfactory to both sides. Valuation based on quoted share prices is impossible since the oil companies operating in the region do not issue shares. Gross cumulative and net cumulative investment would be regarded by the governments as too generous, while, conversely, book value of assets or replacement value of assets would be regarded by the companies as unfairly low. To illustrate, gross cumulative investment in fixed assets in the Middle East at the end of 1970 was estimated by the Chase Manhattan Bank at $7,450 million, and net cumulative investment at $3,685 million, but OPEC estimates of the net book value are: Aramco $400-500 million, Iranian Consortium $400-500 million and Kuwait Oil Company $250 million, suggesting a total of $1,500 million for the four major producers.[10] Lastly, basing compensation on value of oil in place or present discounted value of anticipated profit streams raises even more complex questions, since three of the major variables — the future price of oil, the rate of output and the termination date of the concessions — cannot at present be determined.

The question of termination raises a further point: it seems unlikely that the present Middle Eastern and Libyan concessions will long survive the 1970s although they do not formally lapse until dates ranging between 1994 and 2027.[11] Here, as in the past, Venezuela and Iran may be expected to set the pace. In Venezuela, 70 per cent of the acreage held by companies is due for relinquishment by 1984, and the rest shortly after; the government is already taking steps to ensure that the conditions of expiry will be favorable to it. In Iran, the twenty-five year concession stipulated in the 1954 agreement lapses in 1979. The agreement also provides for three five-year extensions; the companies claim that these are unconditional but the government seems increasingly inclined to contest this interpretation. If Iran should declare the concessions ended by 1979 or, allowing for one extension, 1984, other countries will very likely follow suit.

The continuation of the concessions until the early 1980s is, then, probably the most that can be hoped for. Such an outcome might give time for two processes to work themselves out. First, the consuming countries may by then have found alternative deposits of oil — in the North Sea, Arctic, or Alaska, from shale or from tar-sands — or have developed nuclear or other sources of energy (see Chapter Four). Secondly, the oil producing countries may be sufficiently integrated in the world market and so dependent on oil exports that they may not consider it in their own interests to disrupt it.

To repeat, this is the most hopeful outcome, but it is by no means the most probable as there are powerful forces exogenous to the oil industry which may at any time strike and disrupt it. Indeed such forces have already led to three successful nationalizations: that of the whole oil industry of Algeria in 1971; that of the southern fields of Iraq (Rumaila) in 1967, which are just entering the production stage and for whose output markets are being opened in Ceylon, India and the Soviet bloc (see above); and that of British Petroleum by Libya in 1972.

Forces within the region

The disruptive forces within the region that are likely to impinge on the oil industry may be divided into four main categories: internal revolutions, territorial disputes, national rivalries and ideological antagonisms.

Revolution is endemic in the developing countries, and the oil producers are no exception. It might be thought that in their case oil would act as a lubricant and prevent a social explosion. But the fact is that although rapidly mounting oil revenues have made it possible to achieve great progress and have opened immense prospects, they have also increased the strains on the rather fragile social structures of these countries and heightened the tensions within them. People everywhere tend to ignore the nine successes achieved and concentrate on the one failure. And it is an unfortunate but indisputable fact that, along with rising living standards for the masses, educational advances, welfare services and a huge widening of horizons, oil wealth has greatly enlarged the gap between rich and poor and increased corruption in the ruling circles. As the example of Libya has shown, these two defects may outweigh all the achievements of a regime, and the ensuing discontent can often be harnessed by an ambitious army officer or member of the ruling group. The probability of the overthrow of existing regimes in all the oil producing countries is high, but in one or two — notably Iran — the political and social structure may by now be strong enough to weather the storm. And before a regime is overthrown, it may of course experience several minor coups, such as the recent ones in Oman, Sharjah and Qatar.

Nor does the overthrow of a monarch or sheikh necessarily increase stability. Syria and Iraq had their first revolutions in 1949 and 1958, respectively. They have seen several since and will surely see more, for the discontent that produced the initial revolution is not likely to be allayed by it for long, and the prize of power gets richer every year.

It may be argued that revolution is not necessarily catastrophic for the oil industry, since the new regime will be at least as dependent on its oil revenues as the old. This is true, and companies have found it possible to continue doing business in postrevolutionary Iraq and Libya and to start new enterprises in Algeria. But it is also a fact that these countries have, as mentioned before, carried out significant and successful nationalizations. In other words, whereas the traditional regimes threaten the companies with gradual "appropriation," the revolutionary ones are more likely to resort to expropriation. And for a variety of reasons, expropriation is much more likely to be successful today than it was in Iran in 1951.

Territorial disputes within the region are numerous, but need only brief mention here. Many of these are very old, but all were frozen during the period of British domination of the Persian Gulf and emerged only when it became clear that the British were withdrawing, leaving the local states to fend for themselves. There is Iran's claim to the three tiny islands at the mouth of the Gulf — Abu Musa and the Tumbs — which was enforced by sending occupation troops in November 1971. There is the dispute between Iran and Iraq over the Shatt al-Arab, the river formed by the confluence of the Tigris, Euphrates and Karun, which marks part of the boundary between the two states, constitutes Iraq's outlet to the sea and provides access to Iran's two main ports, Khorramshahr and Abadan. The boundary between Saudi Arabia and Abu Dhabi has given rise to numerous controversies and a few small armed clashes, notably over the Buraimi Oasis. The frontiers between most of the sheikhdoms are not clearly defined, and those between Saudi Arabia, Oman, Yemen and South Yemen cover large areas of contention. Lastly, there is Iraq's claim to Kuwait, which it attempted to enforce in 1961 but was thwarted by the other Arab states, led by Egypt and Saudi Arabia. To these should be added two other disputes outside the Gulf area: the dormant Turkish-Syrian one over Alexandretta and other border areas; and the conflict between Israel and the neighboring Arab states.

National rivalries — as distinguished from territorial disputes between states — exist between the five main ethnic groups inhabiting the Middle East: Arabs, Turks, Iranians, Kurds and Israelis. The Kurds have at various times risen against the Turkish, Iraqi and Iranian governments and will probably do so again; they have thus provided a common interest drawing these three governments together, but at times one or other of them has judged it expedient to encourage Kurdish rebellion in a neighboring state. Since the Arabs gained their independence from the Turks in the First World War, there has been no armed conflict between the two but tensions have been frequent because of border disputes or differing alignments in the Cold War. Arab-Iranian rivalry for hegemony in the Gulf area grew acute once it became clear that Britain was about to withdraw; it has at various times pitted Iran against Egypt and Saudi Arabia, but at present its main antagonist is Iraq. The Arab-Israeli war affects the oil industry only indirectly and will be discussed below.

Lastly there are the ideological antagonisms between the "conservative" and "progressive" states. These terms denote not the internal policies being pursued — for some "conservative" governments have achieved more for their people than has any "progressive" Middle Eastern state — but their attitude towards revolution. The conservative states of the region are the large oil producers — Iran, Kuwait and Saudi Arabia — the small sheikhdoms, Jordan and Lebanon.

On the other side, there are many candidates for leadership. Egypt was for long the fountain of revolution. In the last few years, locked in its fight with Israel, it has paid less attention to other matters, but may do so again — out of either ambition or desperation. Algeria has the prestige of a successful war of national liberation against a major power and of a fairly deep social revolution, but its cultural and geographical remoteness from the center of the action is undoubtedly a handicap. Libya is closer, and has both an enthusiastic leadership and vast sums of money, but its narrow home base hampers effective action. Iraq has the ideology and, potentially, the funds but is crippled by internal instability, and Syria presents the same picture but without the funds. Lastly, there is South Yemen, whose hopeless economy and intractable tribal and other problems have put new life in the long-smouldering rebellion in Dhofar enterprise. South Yemenis, with Chinese arms and advisers, have put new like in the long-smouldering rebellion in Dhofar against the Sultan of Oman, and have pinned down most of his troops and a large part of his resources. The ultimate aim is to overthrow all the present Arab sheikhdoms and form a People's Republic of the Gulf, stretching from Kuwait to Oman and Dhofar. It is clear that in the next few years revolution will not flag for lack of promoters. And it is highly probable that the impact of some of the explosions created by revolutions or border or national disputes will be felt by the oil industry since it is the main source of wealth in the Persian Gulf area, the most tempting prize and the one remaining sector under foreign control.

Forces outside the region

The external forces that impinge on the region, and more particularly on its oil industry, originate in the five world centers of political and economic power: China, the Soviet Union, the United States, Japan and Western Europe.

China's expansion has traditionally lain overland, bringing it close to the Caspian Sea more than once, but it is worth recalling the four great sea expeditions that reached East Africa at the beginning of the fifteenth century. Once again, there has been a resurgence of Chinese interest in the Indian Ocean area, and more particularly East Africa and South Arabia. In addition to the large credits extended to Tanzania and Zambia, China has given small amounts of aid to certain countries in the Middle East and North Africa. In 1954-1970, total credits extended by China were just over $300 million, the beneficiaries being Egypt, South Yemen, Algeria, Sudan, Yemen and Syria.[12] China is also conducting a small but growing volume of trade with the region involving a total turnover (exports plus imports) of $200 million in 1968 and slightly more in 1969. So far this does not include oil, since China has met its very low consumption from its estimated output of 25 million tons a year (500,000 barrels a day); but small quantities may well be sent in the future. China's main impact however, is, likely to be political rather than economic. Unlike the Soviet Union, it is eager to deal with revolutionary movements as well as established states. Arms and experts have been supplied to the guerrilla forces in Dhofar, and contacts have been made with the Palestinian guerrillas, and in all likelihood with other movements. Such relations may prove embarrassing to the Soviet Union — which cannot relish the prospect of being outflanked on the left — and may push it in either of two opposite directions: greater support for the revolutionaries or increased willingness to reach accommodation with the United States.

Soviet involvement in the region is far deeper and more complex. In the first place, its trade has expanded greatly, from a negligible amount in the early 1950s to a total turnover of $1,500 million in 1970, broken down as follows: Egypt $674 million, other North Africa $279 million, Arab Asia $197 million, Iran $257 million, and Turkey $93 million; no trade was reported with Israel. Soviet exchanges with the region accounted for 60 per cent of its total trade with developing countries, and it is now the leading partner of several countries such as Egypt, Syria and Sudan. Judging from the 1969 data, East Europe's trade turnover with these countries in 1970 was over $1,000 million, representing about two-fifths of its total trade with the developing countries.

Secondly, the Soviet Union has supplied large amounts of

economic aid. Credits extended by it in 1954-1970 totalled $3,019 million, broken down as follows: Egypt $1,011 million, other North Africa (mainly Algeria) $418 million, Arab Asia (mainly Iraq and Syria) $652 million, Iran $562 million, and Turkey $376 million. This total represents a little under half of the economic aid granted to all developing countries. It is estimated that just under half of these credits have so far been drawn upon. The Eastern European countries have also extended credits totalling $1,550 million to the region, the main beneficiaries being Egypt, Iran, Iraq and Syria; this represents well over half of East Europe's aid to the developing countries.

Lastly there is Soviet military aid, conservatively estimated for 1955-1970 at: Egypt $2,700 million, Iraq $500 million, Syria $450 million, Algeria $250 million, Iran $110 million, Yemen $100 million, Sudan $60 million, Morocco $20 million, and unspecified sums to Libya and South Yemen. Except for Iran, Libya and Morocco, the Soviet Union is now the principal or sole supplier of arms to these countries. With arms have come military advisers, an estimated 14,000 in Egypt in 1970, 1,000 in Syria and 250 in Iraq.[13]

The expanding trade and economic aid, and still more the military help and diplomatic support given by the Soviet Union, have put it in a predominant position in the main Arab countries — Egypt, Algeria, Iraq, Sudan, Syria and Southern Yemen. Its influence in the other Arab countries is growing. Even in Iran and Turkey it has succeeded in at least partly offsetting the predominance enjoyed by the United States since the Second World War. This process has of course been greatly helped by the increasing power of the Soviet navy in the Mediterranean and the Indian Ocean; although not in a position to challenge the far more powerful American fleet, it does considerably limit the latter's freedom of action. It is also helped by the facilities granted to the

Soviet navy by such countries as Egypt and Syria. The next Soviet policy aim may well be to extend its influence to Persian Gulf oil. For this, there are excellent economic and political reasons.

With regard to economics the Soviet Union has to supply not only its own needs but also those of Eastern Europe and other dependent countries and consumption in both areas is growing rapidly. In addition, oil exports to the West provide it with the foreign exchange required for importing essential machinery and equipment. Finally, Middle Eastern and North African oil and gas can be delivered in many parts of the bloc more cheaply (in real terms, or opportunity costs) than Soviet oil because of the higher production costs or inconvenient location of the latter. Quantification is however rendered difficult by several unknowns. First, as regards Soviet crude production, the target for 1975 was recently set at 500 million tons, a rise over previous figures, and estimates of 625-645 million tons have been mentioned for 1980. These targets imply increases of just over 40 and 80 per cent, respectively, over the 1970 level of 355 million, but both are well in line with the recent growth rate of nearly 8 per cent, compound. Estimates of existing reserves, including the newly developed Tiumen field in western Siberia, suggest that this rate can be sustained without running reserves dangerously low.

Soviet oil consumption has, in the last two decades, lagged behind production. If this trend continues, the margin available for export to the bloc and other regions should widen. But the rate of growth of consumption may accelerate with rising affluence and an increasing tendency to replace coal by oil. Consumption in Eastern Europe is expected to grow rapidly, while production, which is at present negligible, will show little increase. It seems virtually certain that by 1980 Soviet and bloc consumption will either equal or exceed Soviet production, leaving no surplus for export to the West. Moreover, some of the Siberian production may be channelled eastward, e.g., to Japan, and thus not contribute to the needs of European Russia and Eastern Europe. Hence it is not surprising that in recent years Soviet economists have been stressing the burden which their country is carrying as the main provider of raw materials within the bloc, and urging the other countries to increase their investments in the Soviet mining and fuel industries,[14] nor that there are increasing purchases by the Eastern European Countries of Middle Eastern and North African oil. The Soviet Union's own purchases of such oil have been negligible, but it is importing large amounts of gas from Iran and Afghanistan.

The Soviet Union and Eastern Europe should find no difficulty in paying for Middle Eastern and North African oil through barter deals, and it might be expected that their status as purchasers would give them important common interests with the other large oil consumers. But the Soviet's interest in Middle Eastern oil goes far beyond the economic, since they cannot help but recognize it as the Achilles' heel of Western Europe. The dream of turning off the oil faucet and freezing Europe into submission must surely be one of the most persistent in the Kremlin. Though not impossible to fulfill, it will not be easy to achieve. For it implies not only the collapse of Western influence in the region but persuading the Arabs and Iranians that their purposes coincide with those of the Soviet Union, or coercing them into subordinating their interests to its own. Thus, for some years to come, the Soviet Union may be expected to play a more limited and cautious role: backing the oil countries against the companies, siding with the Arabs in their recurring conflicts with Israel and the United States, signing treaties and pacts such as the recent ones with Egypt and Iraq, and taking advantage of Western — and more particularly

American — mistakes which, as in the past, may be expected to be numerous. Soviet influence in the region should continue to grow, but as it does, the Soviet government will be increasingly faced with the kind of dilemmas that have plagued the West and that inevitably arise from the conflicts between the local nationalisms and the various political forces mentioned earlier.

For the last twenty-five years — ever since the Straits and Azerbaijan crises of 1945-1946 — the United States has succeeded in preventing the Soviet Union from dominating the Middle East, protected the southern flank of NATO, and ensured a continued flow of oil to Western Europe. It has also developed economic interests of its own, which though small in relation to the American economy are by no means negligible. First, there are the profits of the American oil companies in the Middle East and Libya, amounting to some $1,400 million a year. Then there is an active and growing trade. In 1969, United States exports to the region amounted to $1,953 million, or 5 per cent of the United States total, and imports of $584, or under 2 per cent. The total turnover of $2,537 million was broken down as follows: Arab countries $1,145 million, Israel $586 million, Iran $439 million, and Turkey $367 million. Part of this was financed by United States aid which, in 1946-1967, amounted to over $8,000 million, broken down as follows: $3,500 million to the Arab countries, $1,000 million to Israel, $1,000 million to Iran and $2,400 million to Turkey. Unlike Soviet credits, all these amounts have been drawn down, as well as some additional sums extended since 1967. To all of this should be added the United States' old cultural and philanthropic ties with the region and its newer, but far stronger ones with Israel.

The United States still holds several positions of strength in and around the region including the Sixth Fleet, the sea and air bases in the northern Mediterranean, the NATO alliance, and the support of powerful local forces. But there is no doubt that its overall influence has greatly diminished. First, willingly or under pressure, the United States and its allies have abandoned practically all their bases in the Middle East and North Africa. Secondly, the Western monopolies — arms, markets, credits, supplies of equipment, sources of cultural exchange — have been broken by the Soviet bloc, and there is also intense competition between the Western States. Thirdly, the upheavals of the last twenty years have, in all the larger Arab countries, replaced friendly, conservative regimes by hostile revolutionary ones. Rightly or wrongly — and sometimes undoubtedly rightly — these regimes see in the United States an enemy intent on overthrowing them, and react accordingly. Fourthly, in the Middle East as elsewhere in the Third World, the United States is the main target of nationalist and "progressive" forces, being debited, by some mysterious process, not only with its own sins but with all the grudges and grievances accumulated under European colonial rule. Fifthly, the prestige of the United States has undoubtedly suffered from its manifest failures in Vietnam and Bangla Desh, and the economic difficulties it is facing. Lastly, and most important, are the repercussions of the Arab-Israeli conflict. This is not the place to go into this complex and tragic question. Suffice it to say that, at present, three things are clear. First, there is no immediate prospect of the Arabs and Israelis coming to an agreement; at best, one can expect latent tension broken by raids and counterraids, and at worst, periodic bouts of warfare. Secondly, whatever its hesitations may have been in the past, since May 1967 the United States has firmly backed Israel, and the Israelis have acted in this assurance. Thirdly, the Arabs have also realized this fact, and consequently have relied increasingly on the Soviet Union, which has given them ample military and diplomatic support. And while one cannot but suppose that neither the United States nor the Soviet Union wants a direct confrontation in the Middle East and must therefore have attempted to restrain their protégés, it would seem that their influence over the latter is limited and that their wishes are ignored when the local state feels that a vital interest is at stake. This combination of circumstances means that, sooner or later, an Arab-Israeli explosion is likely to have repercussions elsewhere. American oil interests present an irresistably attractive prize; and the Arabs are more or less bound to retaliate against the United States by disrupting the oil industry. Another set of repercussions is also likely to be set off by the Palestinian guerrilla raids and the Israeli reprisals. Between them, Palestinians and Israelis will most probably succeed in undermining and eventually overthrowing the existing re-gimes in Jordan and Lebanon, which will in all likelihood be succeeded by radical governments hostile to the West. In other words, everything points to a further crumbling of the United States position in the Middle East and North Africa.

Such a prospect must surely cause some dismay among United States policy makers, but probably nowhere near despair. For, after all, the importance of the Middle East to the United States is not very great. But for Japan and Western Europe, the Middle East and North Africa, and more specifically their oil, are literally vital. The position of Japan can be described quite simply. It carries on a substantial volume of trade with, and, more importantly, draws over 80 per cent of its oil from the region, some of it supplied by Japanese companies operating in the Gulf. These facts and the possibility of substantial aid should eventually give Japan considerable economic leverage, but at the moment it is the dependence that counts. So far, Japan has relied on the United States and Britain to defend its oil interests in the region. Presumably it has no choice but to continue doing so, but one can surmise that the Japanese are making efforts to establish friendly links with the governments of the region and to dissociate their image from that of the United States, as well as to obtain further oil concessions, both in the Persian Gulf area and in other regions, that will decrease their dependence on the oil companies.

The position of Western Europe is, in essence, similar to that of Japan, but the area not being homogeneous, the situation is more complex. Western Europe is the main trading partner of the region, its turnover exceeding that of the United States and the Soviet bloc combined. It is, more particularly, the main oil market, absorbing nearly 50 per cent of Middle Eastern oil and almost the whole of North African production, and relying on these two areas for over 80 per cent of its own needs. But an important distinction must be drawn between Britain, which has large interests as a producer of Middle Eastern oil, and the continental countries, which are primarily consumers, with France playing an intermediate and crucial role.

For 150 years Britain's interests in the Middle East were mainly strategic, and could be fully safeguarded by its navy and small garrisons. But withdrawal from India in 1947 removed the main reason for the British presence in the region, and the Suez expedition of 1956 showed that Britain could no longer pursue an independent policy in the Middle East, even in alliance with France. The recent withdrawal from the Persian Gulf marks the end of an important period in Middle Eastern history.

By now, Britain's interests are almost solely economic. It carries on a large volume of trade with all the Middle Eastern and North African countries and draws the bulk of its oil from the region. Its profits on oil operations are of the order of $700 million a year, a large item in the British balance of payments. Lastly, the sterling balances held by Kuwait and other oil producers run into the hundreds of millions.

At present, Britain has to rely on a combination of United States power and local goodwill to protect its interests. The inadequacy of both was recently demonstrated by Libya's nationalization of British Petroleum's holdings, in retaliation for presumed British connivance with Iran over the seizure of the three Persian Gulf islands (see above). For the time being, Britain has probably no choice but to follow United States policy in the region. However, as shown in the prolonged negotiations concerning the Arab-Israeli conflict, the views of the two countries are not identical. Should Britain finally join the European Economic Community, it may find that it has enough common interests with the other members to help frame a coordinated policy on Middle Eastern oil.

Germany's concern arises from its large volume of trade and, especially, its almost total dependence for oil supplies; its interests as an oil producer in the region are negligible. Until now, compelling overall military and political considerations have made it follow United States policy in the Middle East as elsewhere, but its growing strength, on the one hand, and its needs as an oil consumer on the other, may gradually bring about a change. In this context, Brandt's recent visit to Iran is quite significant.

Italy is in the same position as regards trade and oil and it was also highly dependent on the Suez Canal for navigation. Under the vigorous and buccaneering leadership of Enrico Mattei, it looked at one time as though the state petroleum company, ENI, would try to challenge the position of what he called "The Seven Sisters" in the Middle East and North Africa. But in fact, ENI's production is still small and is likely to remain so, and Italy's interests are predominantly those of an oil consumer. During the last few years, it has tried to cultivate better relations with the Arabs without straying beyond the bounds set by NATO policy.

France shares the trade and oil interests of Italy and Germany, but unlike them it draws significant profits from oil production — about $100 million a year from the Middle East and a similar amount from Algeria. Under De Gaulle, it followed an independent policy in the region, most strikingly revealed by the arms embargo on Israel and the arms deal with Libya. This earned it much Arab applause, and no doubt some goodwill, but it did not prevent the nationalization of the French oil companies in Algeria and it also looks as though ERAP's concession in southern Iraq will be cancelled.[15] The fact remains that, by itself, France has insufficient weight to exercise leverage in the region. But it may well have blazed a path for a West European bloc, of which it might take leadership in this matter. Assuming enough coordination between its four major members, such a bloc would confront the oil producers with tremendous economic power. This could enable it to reach a direct agreement with the oil producing governments, bypassing both the United States government and oil companies and the Soviet Union. Such an agreement might be welcomed by the Arabs and Iranians, since they have no outstanding grievances against Europe, are strongly attracted to its culture, know that it is their largest potential market (for oil as well as other raw materials and foodstuffs) and do not regard it as posing the kind of political and military threat that the United States and the Soviet Union constitute. In the past, the dominant sentiments of the region were Anglophobia or Francophobia, but these have been replaced by anti-American and, increasingly, anti-Russian sentiments.

CHAPTER FOUR

Prospects and Policies

So far, the world has successfully coped with three Middle Eastern oil crises, in 1951, 1956, and 1967. The 1951 nationalization cut off the production of Iranian crude and, what was more important, of refined products from Abadan; the shortage was overcome by increasing output in the other Gulf countries, utilizing excess capacity in European and American refineries, redeploying tankers, and redirecting oil flows. The 1956 crisis arose from the blockage of the Suez Canal and the stoppage of the pipelines through Syria; production remained largely unaffected and the rerouting of tankers around the Cape together with an increased output in the United States and Venezuela made it possible to meet Europe's needs. In 1967, the Canal was again blocked, the flow through the pipelines was temporarily stopped, and a partial boycott was briefly imposed by some of the Arab producers; but the abundance of large tankers, and the vast increase in Libyan output, enabled the companies to meet their obligations. Clearly, today the transit countries, Egypt, Syria, Jordan and Lebanon, no longer pose a real threat. But dependence on Middle Eastern and North African oil has greatly increased since world consumption has vastly expanded, and the available reserve capacity in the United States and Venezuela is far smaller than it was ten or twenty years ago. The events of 1970-1971 have shown how great a pressure the producers, if united, can exert. Means must be found to reduce this dangerous dependence, and these will be considered under two headings, demand and supply.

Demand

The aim here must be to decrease the overall demand for energy, or more precisely, to slow down its rate of growth. This demand is a function of three variables: growth in population, rise in per capita real income, and the "energy coefficient," i.e., the increase in energy consumption accompanying a given rise in per capita income (see Chapter One). There is no need to discuss the first factor at length; most people would, by now, agree that a slowdown in the rate of population growth is desirable, welcome the decline in birth rates that has been observed in Western and Eastern Europe, the United States, the Soviet Union, Japan and, apparently, China, and urge the less developed countries to bring their population growth under control. The question of income is more controversial; some have gone so far − for many reasons of varying degrees of validity − as to advocate Zero Growth. But even for the advanced countries this would pose social problems that might well prove insoluble, and that would outweigh the by no means certain benefits. For the poorer ones it would be a thoroughly inhuman prescription. The most that can be hoped for here is a planned and internationally agreed slowdown, accompanied by a redirection of priorities and efforts for the richer countries, and the difficulties this raises hardly need emphasis.

The difficulty of reducing the energy coefficient is only slightly less, but the need to do so is compelling. First, there is the fact that known energy supplies are not unlimited, and must therefore be husbanded — though this statement may be refuted by further developments in nuclear energy. Secondly, and of more immediate concern, is the increasingly serious environmental pollution caused by excessive use of energy in advanced countries.

One way of reducing consumption in these countries is relatively easy, the elimination of the enormous waste of energy, particularly in the United States.[1] Houses are overheated in winter and overcooled in summer, almost certainly with bad effects on health. Lights are left blazing in empty hotel rooms and other places. Open refrigeration shelves in supermarkets chill the atmosphere. Pilot lights burn enormous guantities of gas. Reduction of such waste would save appreciable amounts of energy without causing real inconvenience, much less hardship; but it would demand a profound change in attitudes which are based on the implicit assumption that energy is almost costless and inexhaustable. It is therefore very gratifying to note such symptoms of awareness as the current campaign to "save a watt." A further source of saving would be changes in building materials and methods; to allow for better insulation and more natural and less artificial means of ventilation and cooling.

Still greater savings could be achieved in the field of transport. The present standard American car, weighing some 4,000 lbs. and with a horsepower of about 230 is just too great a consumer of energy, especially when driven at the high speeds of which it is capable; the corresponding figures for a Volkswagen "Beetle" are 1,800 lbs. and 60 h.p. In 1948, James Forrestal stated (not without anxiety) that "unless we had access to Middle Eastern oil, American motorcar companies would have to design a four-cylinder motorcar sometime within the next five years."[2] It is time this suggestion was taken up. And, needless to say, much greater efforts must be made to develop mass transit and decrease the growing Western dependence on the plane and private automobile. Of course, such policies would meet with enormous resistance, but here too there is an encouraging, and indeed dramatic, sign in the recent rejection of the SST plane.

Still more difficult would be some other suggested changes, which are highly desirable not only to conserve energy but for a variety of social reasons. These aim at breaking up the existing huge urban sprawls with commuters travelling vast distances to and from their work, and replacing them with smaller, self-contained units, where people would live near their place of business and travel much less for work or play. But this borders on the Utopian, and it is time to turn to questions of supply.[3]

Supply

The measures relating to supply fall into two groups: long-run and short-run measures. Among the former are additional supplies of coal and nuclear power. Little need be said on either in this paper, except that the output of coal is expected to double by 1985 in the United States but to decline in Europe (see Chapter One). Perhaps more emphasis should be placed on strip-mining, which is both cheaper and far less dangerous to the miners. At present strip-mining is under attack because of its ecological effects, but regulations obliging the companies to rehabilitate the mined areas should not be impossible to enforce. Output of nuclear power is expected to increase very rapidly in the 1970s and 1980s. So far, the industry has not fulfilled the high hopes set on it in the last two decades. Capital requirements have proved higher than expected; the cost per unit of electricity produced has been above that derived from fossil fuels; and the need to safeguard against radiation perils has permitted much obstruction, resulting in delays and higher costs. Clearly, a greater national commitment is required, and President Nixon's request to Congress in June 1971, for $2,000 million over the next decade for development of a commercial fast breeder reactor is a step in the right direction.

Coal and nuclear power can fill part of the anticipated increase in energy requirements, but do not meet the need for liquid fuels. However, there are three other resources, which are abundant in North America, and which could, in principle, do so: shale, tar sands and coal for hydrogenation.[4]

Oil-bearing shale rocks are found in huge deposits in Colorado, Utah and Wyoming, accounting for about half the world's total. Their estimated oil content is around 600 billion barrels, an amount equal to the whole of the world's proved natural petroleum reserves (see Table 4). The technological problems involved are complex, arising from the need to crush huge quantities of rock, heat it to retort the oil, and refine the oil to bring it up to normal crude petroleum standards; but by and large they have been mastered. Many other difficulties remain, however, including ecological ones, in view of the immense amounts of rock that have to be processed and disposed of, and financial difficulties, due to the huge investment required. Capital cost per barrel/day of capacity has been variously estimated at $2,550 to $4,000, which implies outlays of several billions of dollars if a substantial output is to be obtained. Various government and business estimates have suggested that by 1972 shale oil could be produced at about $3 a barrel, and by 1976 at perhaps as little as $2.15, which would be competitive with domestic petroleum. For the foreseeable future, the upper limit of shale oil production has been put as high as 2 million barrels a day (100 million tons a year), but

Schurr and Homan doubt that actual capacity by 1980 will exceed 350,000 barrels and judge 200,000 a more likely figure. In the meantime, much can be done to expedite development, including not only accelerated research and subsidization of pilot plants but adoption by the Department of the Interior of a clear policy on the leasing of the public lands in which the bulk of the shale deposits are located.

Tar sands, like shale, contain petroleum that cannot be recovered in the normal way but requires special extraction in this case by washing, thinning and processing to reduce viscosity and bring it up to normal standards. Deposits are found in various parts of the world including the United States, but the largest concentration is that of Athabasca in Alberta, Canada, whose contents have "been estimated at 285 billion barrels, with 85 billion considered currently minable." Here too, the technical difficulties have been overcome, and a plant with a capacity of 45,000 barrels a day of crude has been operating since 1967. But again, the financial and economic prospects are not encouraging. The plant cost a reported $235 million, or about $5,000 per barrel/day capacity, and initial production costs, at about $2.75 a barrel, have been higher than anticipated. Schurr and Homan suggest an upper limit of 300,000 barrels a day by 1980, and judge 150,000 more likely.

In Germany during the Second World War, substantial amounts of oil were derived from coal by hydrogenation, and for several years a commercial plant has been functioning in South Africa. The advantages of this over the two preceding methods are first, that less materials have to be handled and disposed of, since the amount of coal required to produce a barrel of oil is much smaller than that of shale or tar sand, and secondly, that in the United States, Western Europe and elsewhere, coal reserves are available near consuming centers.

But capital costs are high — about $5,000 per barrel/day capacity — and with present costs of coal and hydrogen, this method is not commercially attractive. Much remains to be done to improve technology, and here too government help in greater amounts is needed. If the recently completed pilot plant in the United States shows satisfactory results, it may be possible to produce 300,000-500,000 barrels a day by 1980.

In other words, output of oil from shale, tar sands and coal may, by the end of this decade, amount to some 600,000-800,000 barrels a day (30-40 million tons a year) — though the figures may be higher if oil prices continue to rise. This is a tiny quantity in relation to United States — let alone world — consumption, and the vast investment required would be justified in terms of insurance against interruption of outside supplies and promise of increasing future output rather than because of its actual contribution.

More fruitful is the search for new sources of oil outside the Middle East. The last decade has seen some very important discoveries, most notably in North Africa but also in Nigeria, which has become a big producer. Large deposits (some 5 billion barrels each) have been located in Ecuador and Peru, and one may expect both countries to make significant contributions in the years ahead. Output in Indonesia has been increasing steadily, and its offshore areas, as well as parts of the southern China Sea, hold great promise. The North Sea deposits, estimated at 3 billion barrels with hopes of more, are already supplying Western Europe with large amounts of natural gas, and their planned oil output for 1975 is 250,000 barrels a day in the British sector and 300,000 in the Norwegian, with further increases expected later.[5] Lastly, there are the Alaskan fields, whose reserves may be 30 billion barrels or over, and those of the Canadian Arctic, which may be twice as great. Clearly, the development of these resources would put the world in a better position to cope with possible interruptions in supplies from the Middle East and North Africa.

Another very necessary measure is a change in United States oil and gas policy. Government regulation has resulted in natural gas prices that are too low relative to other fuels: thus "crude oil" at the wellhead in the United States sells for over 50 cents per million B.t.u., compared with about 15 cents for gas at the wellhead and about 18 cents for coal at the mine mouth.[6] This seems to have discouraged development and resulted in a shortage of natural gas; clearly, some increase in prices seems warranted, and decontrol should be seriously considered.

But if the gas industry has been hurt by government regulation, the oil industry has been favored by the combination of quotas and prorationing (see Chapter One) to the extent of about $3,000 million a year. This has imposed not only a burden on consumers but also a severe handicap on industries, such as the chemical, which use large amounts of oil as a fuel or raw material.[7] Two moves are indicated, both of which would encounter immense political difficulties. First, every effort must be made to "unitize" production in as many fields as possible, since this would both reduce costs of production and increase the amount of oil that could ultimately be recovered from the fields. Secondly, the present quotas on imported oil must be greatly liberalized. The main objection to this is that it would increase the dependence of the United States on outside — more specifically Middle Eastern and North African — sources, and this consideration must be taken very seriously. As against that, increased imports would mean a considerable lowering in oil prices. More important, they would make it possible to conserve American oil for use in future emergencies, in the best and cheapest way possible —

underground. And if such a measure were coupled with subsidization of exploration, the other main objection — viz., that lower prices would discourage exploration and therefore reduce new discoveries well below potential — would also lose its force.

All these are long-run measures, but others are also being taken to cope with short and unexpected interruptions in supply. The most important is the building of a large tanker fleet, since any interruption, by dislocating existing patterns of supply, lengthens average hauls and necessitates a larger volume of shipping. This may require, at times of excessive tonnage, the laying up of part of the tanker fleet in "mothballs," with government help. The second measure is stockpiling by consuming countries. "During the 1970 crisis, Europe had only some 60 to 65 days supply at hand (which has now been raised to a target of 90 days); Japan and the developing countries had even less. Realistically even a 90-day stockpile begins to lose its assurance after 20 to 30 days of withdrawal have taken place."[8] Of course, such stockpiling is extremely expensive, running at about $2.5 to $3 billion for Western Europe alone for 60 days consumption at the anticipated 1975 level.[9] But, as Levy points out, "experience has shown that the cost of exposure could be even greater."

Long-run measures to decelerate the increase in energy demand and to develop additional sources of energy and oil may somewhat lessen overall dependence on Middle Eastern and North African oil supplies. Short-run measures such as stockpiling may enable consumers to meet sudden disruptions in the oil industry of one or more countries. But, for the next decade or two, there does not seem to be a real, durable, alternative to Middle Eastern and North African oil, and somehow or other, the world, including the inhabitants of the region, will have to adjust itself to that fact.

Under the best circumstances this will not be easy, but it may help reduce the problem to its true proportions if some distinctions are made. First there is the question of Middle Eastern and Northern African oil prices. All indications are that these will rise, steadily and appreciably, over the next decade or two. In itself, such a development is not catastrophic, since the main importers (Western Europe, Japan and, increasingly, the United States) are rich and since the prices of their exports to the oil producing countries are also rising steadily. Moreover, an increase in oil prices may serve a very useful economic function by forcing an overdue reduction in consumption, stimulating exploration for oil in other regions,and encouraging the development of alternative sources of energy. Compared to costs of production, the price of Middle Eastern and North African crude oil has been very high. But compared to the price consumers are presently paying for oil products, including the taxes levied by their own governments, it has been low. And, taking a broad social point of view, for a commodity which may be exhausted within a few decades, oil is at present being provided far too cheaply.

The second aspect pertains to the ownership of the agencies producing and selling oil and to their earnings. For the reasons given above, it is unlikely that the present pattern of concessions will long outlast the 1970s (see Chapter Three). Yet the large oil companies may still have an important, and moderately profitable, part to play as contract agents in exploration and production and as purchasers, transporters, refiners and marketers of Middle Eastern and North African oil.

Lastly, there is the most basic question of all — the assurance of continued flow of Middle Eastern and North African oil to its present markets. No firm predictions can be made on this subject. It may be hoped that, as the income accruing to the producing countries gets bigger and bigger, and their economies become increasingly dependent upon it, enlightened self-interest will restrain them from disrupting the flow of oil. More precisely, one can hope that even if one or two countries are swept by gusts of political passion, a sufficient number will continue to be influenced by economic considerations and not allow themselves to be blown off course. The likelihood of total disruption can also be reduced by the short-run precautionary measures discussed earlier, And, as long as oil continues to flow to its markets, all other issues are minor. Of course, a take-over of the region by the Soviet Union would completely change the picture, but such an eventuality can hardly be investigated within the limits of this study.

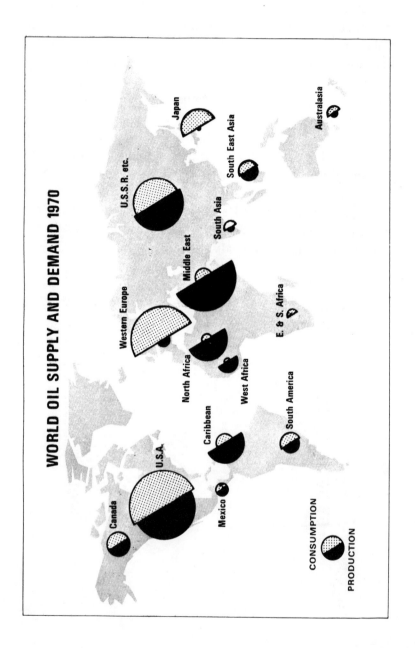

WORLD OIL SUPPLY AND DEMAND 1970

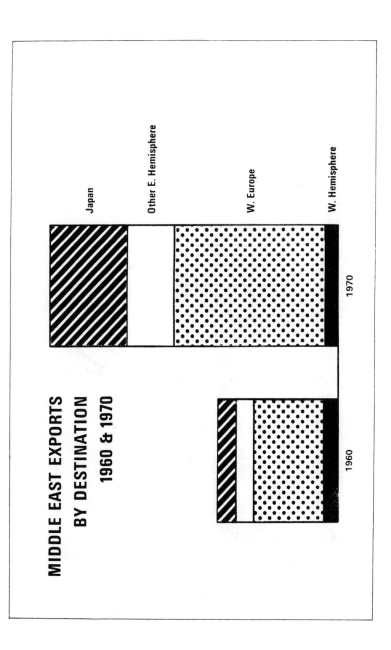

MIDDLE EAST EXPORTS
BY DESTINATION
1960 & 1970

Japan

Other E. Hemisphere

W. Europe

W. Hemisphere

1970

1960

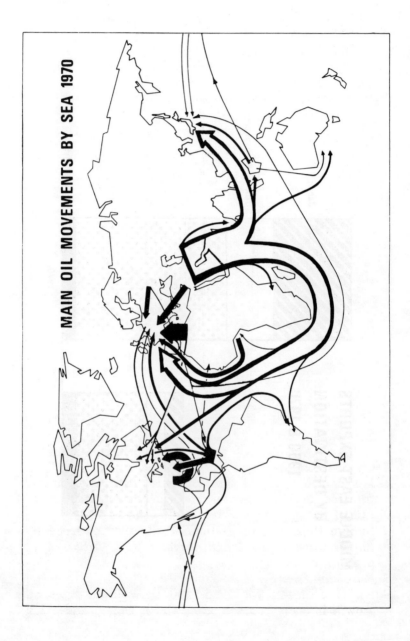

MAIN OIL MOVEMENTS BY SEA 1970

NOTES

Chapter One

1. A. R. Ubbelohde, *Man and Energy* (London: Penguin, 1963), p. 51; *Scientific American* "Special Issue on Energy and Power" 224 (September 1971):37-38.
2. Carlo Cipolla, *The Economic History of World Population* (London: Penguin, 1964), p. 46.
3. David S. Landes, "Technological Change and Development in Western Europe, 1750-1914," *Cambridge Economic History* 6, pt. 1.
4. Joel Darmstadter *et al., Energy in the World Economy* (Baltimore: Johns Hopkins, 1971); British Petroleum Company, *BP Statistical Review of the World Oil Industry.*
5. Darmstadter, *op.cit.,* pp. 20-32.
6. For details, see *ibid.,* pp. 32-40.
7. *Scientific American* 224 (1971):137-140.
8. National Academy of Sciences and National Research Council, *Resources and Man: A Study and Recommendations by the Committee on Resources and Man* (San Francisco, W. H. Freeman, 1969), pp. 201-205. The figures on reserves were taken from the *United States Geological Survey Bulletin;* see also *Scientific American* 224 (1971):63-70.
9. U.S. Congress, Senate Subcommittee on Antitrust and Monopoly of the Committee on the Judiciary, *Hearings,* 91st Cong., 2d sess., 1970, p. 695.
10. National Academy of Sciences and National Research Council, Resources and Man, pp. 168-196.
11. Mohammed Yeganeh, paper presented to United Nations Inter-Regional Seminar on the Development and Utilization of Natural Gas, Moscow, 12-28 October 1971.
12. Interim Report of the National Petroleum Council, *U.S. Energy Outlook: An Initial Appraisal, 1971-1985,* vol. 2, p. xxiii.
13. *Petroleum Press Service,* January 1972.
14. Earlier estimates for 1965-1980 show lower rates of growth in energy consumption, viz. 1, United States 3.5

per cent, Western Europe 4.0, Japan 7.9, USSR 6.5, Latin America 7.4, Africa 6.5, Middle East 9.4, Communist Asia 7.6, and Other Asia 8.2. See Sam Schurr and Paul Homan, *Middle Eastern Oil and the Western World* (New York: American Elsevier, 1971), p. 172.

15. A recent study by the Chase Manhattan Bank suggests that by 1985 the U.S. might have to import 50 per cent or more of its oil requirements for an outlay of more than $25 billion (*New York Times,* 30 June 1972).

16. For details, see Charles Issawi and Mohammed Yeganeh, *The Economics of Middle Eastern Oil* (New York: Praeger, 1962), p. 61.

17. Chase Manhattan Bank, *Capital Investments of the World Petroleum Industry,* 1970.

18. *Ibid.* and Petroleum Department, First National City Bank, *Energy Memo,* October 1971, by Edward Symonds.

Chapter Two

1. Paul G. Bradley, *The Economics of Crude Petroleum Production* (Amsterdam: North Holland, 1967), p. 102; M.A. Adelman, cited in *Petroleum Press Service,* May 1966.

2. Senate Subcommittee on Antitrust and Monopoly, *Hearings,* p. 769; *Oil and Gas Journal,* 13 January 1969.

3. Since the 1930s the situation has been gradually improving, thanks to increasing government regulation and the growing practice of "unitization." One example will suffice: "In the early stages [i.e., 1930] the East Texas field had over 1,700 operators and nearly 3,000 separate leases with diverse mineral interests," leading to a chaotic situation, the collapse of oil prices and the calling of state troops to enforce conservation regulations. "Without regulation, the East Texas field would be essentially depleted today. Instead, in 1968 the field produced 53 million barrels and has an estimated reserve producing capacity considerably above that" (Jim C. Langdon, Chairman of Texas Railroad Commission,

81

Senate, Subcommittee on Antitrust and Monopoly, *Hearings*, p. 677).

4. *Petroleum Press Service*, September 1970.
5. Arabian American Oil Company, *Middle East Oil Developments* (New York, 1952), p. 9.
6. Issawi and Yeganeh, *The Economics of Middle Eastern Oil*, p. 53. In 1970, only 96 development wells (and 88 exploratory wells) were completed in the Middle East; the corresponding figure for Venezuela was 492, and for the United States, 12,230. In Libya, 164 development wells (and 86 exploratory), and in Algeria, 73 (and 42 exploratory) were completed (*Oil and Gas Journal*, 18 October 1971).
7. Newton, Senate, Subcommittee on Antitrust and Monopoly, *Hearings*, pp. 41-77; the following account draws heavily on this excellent paper.
8. All these figures refer to building costs outside the United States. In this country costs are higher, being estimated by the Maritime Administration in 1971 at $409 dwt for a 50,000 tonner, $231 for a 200,000 tonner and $178 for a 400,000 tonner. Recently, it has been estimated that costs per ton of oil transported on 100,000 dwt ships were only 40 per cent of those on 25,000-ton ships, and only 25 per cent of those on 300,000 tonners (*The Economist*, 5 February 1971).
9. *Petroleum Press Service*, June 1970.
10. For a thorough analysis, see B. Hansen and K. Tourk, "An Economic Appraisal of the Suez Canal Project," Working Papers in Economics, Department of Economics, University of California at Berkeley, October 1971.
11. Newton, in Senate Subcommittee on Antitrust and Monopoly, *Hearings*.
12. *Ibid.*
13. M. A. Adelman, in Senate Subcommittee on Antitrust and Monopoly, *Hearings*, and various other publications.
14. *Petroleum Press Service*, December 1971; *International Financial News Survey*, 9 February 1972.
15. The original Iranian concession, which had provided for payment of 16 per cent of net profits, was changed in

1933 at the government's request to a 4-shilling royalty.

16. See Issawi and Yeganeh, *The Economics of Middle Eastern Oil*, p. 114.
17. *Ibid.*, table, p. 109.
18. For fuller details, see Schurr and Homan, *Middle Eastern Oil and the Western World*, p. 122.
19. *Petroleum Press Service*, September 1971.
20. On all these questions, see Schurr and Homan, *Middle Eastern Oil and the Western World*, chapter 10, which gives a good and detailed account.
21. For a list, see *Petroleum Press Service*, March 1970.
22. *New York Times*, 5 March 1972. An important development has been announced still more recently, i.e., the arrival at the Iraqi port of Fao of oil from the North Rumaila field. This field was part of the area expropriated from the operating companies (IPC Group) in 1961 and developed since 1969 by the Iraq National Oil Company with Soviet help. The oil was reported ready for loading on Soviet and Iraqi tankers (*New York Times*, 7 April 1972).
23. Speech by President Boumediene (*Arab Oil and Gas*, 16 November 1971); oil tax revenues alone will rise from $280 million to $640 million.
24. Schurr and Homan, *Middle Eastern Oil and the Western World*, chapter 8; their study covers Iran, Iraq, Kuwait, Libya and Saudi Arabia.
25. Foreign exchange holdings are shown in *International Financial Statistics;* GNP figures have been computed from a variety of national and international sources.
26. *Petroleum Press Service*, September 1970.

Chapter Three

1. Issawi and Yeganeh, *The Economics of Middle Eastern Oil*, pp. 172-173.
2. Charles Issawi, "Coming Changes in the World Oil Industry," *Midway*, Summer 1968.
3. John Vafai, "Conflict Resolution in the International Petroleum Industry," *Journal of World Trade Law*, July/August 1971.

4. Walter Levy, "Oil Power," *Foreign Affairs,* July 1971.
5. First National City Bank, *Energy Memo,* October 1971.
6. The figures have been derived as follows: in 1970, the Seven Companies accounted for 81 per cent of eastern hemisphere crude production (14.7 billion barrels a day out of 18 billion, see *ibid.,* January 1972). Their payments to eastern hemisphere governments amounted to $4,886 million, out of an estimated total of just under $6,000 million, or again, 80 per cent. Assuming, therefore, that their earnings were also 80 per cent of total earnings in the eastern hemisphere, would put the latter at $2,350 million.

In 1970 payments by all oil companies to Middle Eastern governments amounted to $4,189 million (*Petroleum Press Service,* September 1971), i.e., these governments received 70 per cent of total payments by all companies in the eastern hemisphere. Assuming that the ratio of earnings to payments was the same within the Middle East and in other parts of the eastern hemisphere, all oil company earnings in the Middle East would be 70 per cent of $2,350 million or $1,645 million. A rough check of this figure is to assume that earnings per barrel on total production in the Middle East were equal to the eastern hemisphere average of the Seven Companies, i.e., 33 cents. For the Middle East's total production of 5,057 million barrels, this would give a figure of $1,685 million. For Libya's production of 1,212 million barrels, it would give a figure of $403 million.

At the end of 1970, total net investment in fixed assets in the Middle East was put at $3,685 million (Chase Manhattan Bank, *Capital Investments of the World Petroleum Industry,* 1970). The addition of liquid assets may raise the total to around $4,000 million. In Libya, net investments were a little above $1,000 million, suggesting a similar rate of return. It should be noted that profits refer to production alone, whereas the investment figures include refining and other activities; in other words, the rate of profit is somewhat understated.

7. On the subject of participation, see a lecture delivered by Edith Penrose in Algiers (*Arab Oil and Gas,* 1 November 1971).
8. *Ibid.,* 1 October 1971; *New York Times,* 23 January 1972.
9. *Arab Oil and Gas,* 1 October 1971.
10. *Ibid.*
11. For a list, see *Petroleum Press Service,* December 1971.
12. U.S. Department of State, Bureau of Intelligence and Research, "Communist States and Developing Countries: Aid and Trade in 1970," 22 September 1971.
13. *Ibid.*
14. An article in *Voprosy Ekonomiki,* December 1971, points out that although the Soviet Union produces three-quarters of the industrial output and 60 per cent of the agricultural output of Comecon, it makes 80 per cent of the capital investments, largely because it is the major raw materials producer and urges increased investment by the other members in Soviet raw materials production.
15. *Arab Oil and Gas,* 1 December 1971.

Chapter Four

1. Needless to say, there is also much waste in the production and transmission of electric and other forms of energy through inefficiency. Industry is, of course, aware of this and is continuously trying to reduce it, but more research may be warranted.
2. James Forrestal, *The Forrestal Diaries,* ed. Walter Millis and E. S. Duffield (New York: Viking, 1951), p. 357.
3. It should also be noted that many of the measures designed to reduce pollution require an increased use of energy. For example, the average gasoline mileage for 1971 model American passenger cars was 6.6 per cent below the figure for 1970 because of changes in design to control exhaust emissions, and the 1972 model will be 6.2 per cent below the 1971 (*Oil and Gas Journal,* 24 January 1972). Cleaning the air and waters will also require enormous amounts of energy.

4. The following account draws heavily on the excellent analysis in Schurr and Homan, *Middle Eastern Oil and the Western World*, chapter 4; see also National Academy of Sciences and National Research Council, *Resources and Man*, pp. 196-200.
5. A target of 1,500,000 barrels per day has been mentioned for Britain in 1980 (*New York Times*, 16 March 1972).
6. Edward Symonds, in Senate, Subcommittee on Antitrust and Monopoly, *Hearings*, p. 841.
7. For an excellent discussion, see the testimony of Adams, Adelman, Lichtblau and others in *Hearings*, pt. 1.
8. Levy, "Oil Power," *Foreign Affairs*, July 1971.
9. For details, see Schurr and Homan, *Middle Eastern Oil and the Western World*, pp. 80-82.

86

SELECT BIBLIOGRAPHY

Books

Bradley, Paul G. *The Economics of Crude Petroleum Production.* Amsterdam: North Holland Publishing Co., 1967.

Darmstadter, Joel *et al., Energy in the World Economy.* Baltimore: Johns Hopkins, 1971.

Issawi, Charles and Yeganeh, Mohammed. *The Economics of Middle Eastern Oil.* New York: Praeger, 1962.

National Academy of Sciences and National Research Council. *Resources and Man: A Study and Recommendations by the Committee on Resources and Man.* San Francisco: W. H. Freeman, 1969.

Schurr, Sam and Homan, Paul. *Middle Eastern Oil and the Western World.* New York: American Elsevier, 1971.

Periodicals.

British Petroleum Company. *BP Statistical Review of the World Oil Industry.* Various annual issues.

Chase Manhattan Bank. *Capital Investments of the World Petroleum Industry.*

Levy, Walter. "Oil Power." *Foreign Affairs,* July 1971.

National Petroleum Council. *US Energy Outlook,* November 1971.

Scientific American. "Special Issue on Energy and Power," September 1971.

Documents

U.S. Congress, Senate, Subcommittee on Antitrust and Monopoly of the Committee on the Judiciary, *Hearings,* 91st Cong., 2d sess., 1970.